Nature's Calendar

D1431843

NATURE'S CALENDAR

A Year in the Life of a Wildlife Sanctuary

To Daphne,

COLIN REES

With best wishes,

Colin

JOHNS HOPKINS UNIVERSITY PRESS

Baltimore

© 2019 Johns Hopkins University Press
All rights reserved. Published 2019
Printed in the United States of America on acid-free paper
9 8 7 6 5 4 3 2 1

Johns Hopkins University Press
2715 North Charles Street
Baltimore, Maryland 21218–4363
www.press.jhu.edu

Library of Congress Cataloging-in-Publication Data

Names: Rees, Colin P., author.
Title: Nature's calendar : a year in the life of a wildlife sanctuary / Colin
    Rees.
Description: Baltimore : Johns Hopkins University Press, 2019. | Includes
    bibliographical references and index.
Identifiers: LCCN 2018029958 | ISBN 9781421427430 (paperback : alk.
    paper) | ISBN 1421427435 (paperback : alk. paper) | ISBN 9781421427447
    (electronic) | ISBN 1421427443 (electronic)
Subjects: LCSH: Wetland ecology—Maryland.
Classification: LCC QH541.5.M3 R44 2019 | DDC 577.68—dc23
LC record available at https://lccn.loc.gov/2018029958

A catalog record for this book is available from the British Library.

Special discounts are available for bulk purchases of this book. For more informa-
tion, please contact Special Sales at 410-516-6936 or specialsales@press.jhu.edu.

Johns Hopkins University Press uses environmentally friendly book materials, including recycled text paper that is composed of at least 30 percent
post-consumer waste, whenever possible.

*To all who have contributed to the protection of Jug Bay
and to Winston, Carolina, and Virginia*

Contents

Foreword

With this book, the author convincingly portrays the passing of the seasons in a sanctuary that forms part of a regional parks complex of Anne Arundel County in Maryland's Department of Recreation and Parks, which I feel fortunate to oversee. Although I am familiar with Jug Bay's unique landscape, I was often surprised and intrigued as I turned the book's pages. The author not only instructs us in the workings of climate change but also spotlights the life and times of the "lowest" creatures. I came away richer through my reading and have gained a greater insight and appreciation for this national ecological treasure.

While Colin Rees paints with a broad brush, the details are aplenty as he explains the surprising germinations during winter, strategies to compete successfully in the emerging spring, how vegetation shapes summer's lease, and the inexorable slide into autumn, with its provisioning for winter. He portrays a world of remarkable complexity and beauty.

Many have praised the unique qualities of the wetlands and their associated watersheds and have fought for their protection for economic as well as ecological reasons. The reader will have no problem in appreciating the increasing impacts of climate change on the wetlands and their supportive infrastructure. Consequently, what is revealed at Jug Bay can be scaled up to state and national levels and is an invitation for us to take action, although challenging, to help preserve and protect these wetlands.

By combining science with poetry and luminous prose, this book offers a new world to be entered, enjoyed, and to feel passionate about.

Rick Anthony
Director, Recreation and Parks,
Anne Arundel County, Maryland

Acknowledgments

It is a great pleasure to acknowledge the many who made contributions to this book. Judy Burke, Barbara Johnson, Chris Swarth, and Claire Trazenfeld provided valuable reviews and comments. Pati Delgado and her staff (Liana Vitali, Debra Gage, Melinda Fegler, and Diane Benedetti) at the Jug Bay Wetlands Sanctuary gave constant support, as did some of the Sanctuary's dedicated volunteers (Cynthia Bravo, Chuck Hatcher, and Mike Quinlan).

I am much indebted to Frode Jacobsen, Rob McEachern, Kerry Wixted, Jim Brighton, Bill Hubick, Debra Gage, Nancy Martin, Bill Harms, Robert Ferraro, Richard Orr, Gary Van Vlesir, James Parnell, Chris Swarth, and Giles Gonthier for the use of their exceptional photographs (many drawn from the Maryland Biodiversity Project), as well as to the Maryland-National Capital Park and Planning Commission, the Jug Bay Wetlands Sanctuary, and Dave Linthicum, for producing the map of the Jug Bay area.

Tiffany Gasbarrini, Lauren Straley, Kyle Kretzer, and Kathleen Capels provided much help and advice in the production of the book, for which I am most grateful.

Lastly, I have to thank my wife, Valerie, for her constant support and for once again tolerating prolonged periods at the kitchen table when I should have retired to my study. As always, she brought grace and good humor to the enterprise.

Nature's Calendar

Prologue

An interest in nature leads you into a kind of enchanted labyrinth. You
wander from corridor to corridor; one interest leads to another interest; one
discovery to another discovery. It matters little where you begin.

Edwin Way Teale, *Circle of the Seasons*

The months of the year, from January up to June, are a geometric progression
in the abundance of distractions.

Aldo Leopold, *A Sand County Almanac*

Exploding flocks of waterfowl rising from frozen marshland, the arrival of ospreys,
emerging skunk cabbages, spatterdocks, and wild rice, salamanders creeping across
the forest floor to vernal pools, chorusing frogs, spawning fish, migrating warblers
crowding budding trees, turtles sunning on floating logs, the ecological engineering
of beavers—these are but a few events evoking the sights, sounds, and colors of
passing seasons observed and recorded over a year (2017) at the Jug Bay Wetlands
Sanctuary and its environs in Southern Maryland.

This book is a diary of events encountered in which I discovered the joys and
delights at the Sanctuary: some shared, some solitary, some familiar, some less so.
Throughout, I have tried to convey the sweep and drama of the changing seasons,
their moods, the ordinary and the strange—all part of nature's annual cycle, its
continuity. We live in anticipation of the first appearance of a particular plant or
animal, now made more challenging by the effects of climate change.

I often went out with the intent of finding something in particular, be it a flower,
butterfly, bird, or vista. Though my eyes were sharpened by such expectation and
purpose, I invariably found something else just as compelling. I walked the same
trails—marsh edges, woodland, and meadowland—and felt a growing familiarity,
yet also constant surprise and wonder. Songposts, otter scent mounds, dying trees,

or swelling or contracting pools were my markers, but many were my distractions, for to take the same trail is not to see the same view. Rather, it is to notice the slow rate of natural change and chance upon memorable encounters with an otter, the first osprey chick of the season, or monarch butterflies heading for their winter retreat.

This book grew out of an earlier one I had cowritten on the changing seasons experienced by birdlife on either side of the North Atlantic. Upon its completion, I felt the strong need to place birds in the context of all wildlife, using my knowledge as an ornithologist as an entry point. Having been educated in the natural sciences, I was confident I had a head start. In this I was mistaken, and I frequently found myself on a steep learning curve. My salvation was to read the works of the early naturalists, who, by dint of acute observation, made discoveries at almost every step. Without our technological advantage, they had to look harder, be infinitely more patient, and record with a freshness our modern world lacks. Their insights and diligence provided a constant guide.

What follows is an attempt to portray the beauty and wildlife of an exceptional part of Maryland (indeed, of the Eastern Seaboard of the United States), during a yearlong progression of the seasons and the activities of the Sanctuary and nearby areas. It offers the curious and the more serious student of the natural sciences insights into the ecological and behavioral dynamics of its changing landscapes: the infinite feats of biological inventiveness in the face of harsh winter conditions, the annual arrival and departure of migrants, the quickening and sudden emergence of aquatic vegetation, the consuming fires of autumn, and ecological mismatches and species invasions. Above all, it celebrates the richness and complexity of the biological diversity of a continuum: from wetland, upland forest, and grassland during icy to high summer days, with extremes ranging normally from 29°F in January to 89°F in July. Average annual rainfall is 42 inches, with some 26 days of snow.

Birders, herpetologists, botanists, and other specialists have a particular view of the world, and this book invites them to take a broader perspective, crossing disciplinary boundaries, and to marvel at the wholeness and intricate workings of evolutionary processes. No observer of wildlife is bound by the limitations of physical borders or intellectual boundaries; indeed, they can let their imagination soar.

Lastly, in a hectic and often-disturbing world, it is hoped that the reader may find solace in the unhurried rhythms of nature, the changing colors of the seasons, and the wonder of it all. It enjoined me to appreciate nature the more and to accept its moral depths:

> to recognize
>
> In nature and the language of the sense,
>
> The anchor of my purest thoughts, the nurse,
>
> The guide, the guardian of my heart, and soul
>
> Of all my moral being.

William Wordsworth, "Lines Written a Few Miles above Tintern Abbey"

The Setting

Before plunging into an account of the passing seasons, readers may wish to appreciate something of the wider context of the Jug Bay Wetlands Sanctuary, as well as its major features. As part of the Chesapeake Bay watershed, the Sanctuary houses unique natural habitats and has been at the crossroads of history, both ancient and modern.[1] It provides a living laboratory for everyone with an interest in the natural world.

Located along the middle section of the 110-mile Patuxent River, the Sanctuary lies in the lower coastal plain of Maryland and, in turn, within the Chesapeake Bay and the Patuxent River watersheds. The Sanctuary was established in 1985 and is operated by the Anne Arundel County Department of Recreation and Parks. Three streams—Two Run Branch, Pindell Branch, and Galloway Creek—flow through the Sanctuary and into the Patuxent River. It includes more than 1,700 acres of open water, wetlands, vernal pools, forests, and meadows housing many invertebrates, fish, amphibians, reptiles, birds, and mammals, as well as some 15 miles of trails and boardwalks offering glimpses of the region's rich history. On the southern edge of the Sanctuary, the River Farm includes 12 acres of agricultural fields, where habitat management and organic farming are practiced.

In 1990, the Sanctuary became a component of the Chesapeake Bay National Estuarine Research Reserve System and was subsequently recognized as a "nationally important bird area" by the American Bird Conservancy and the National Audubon Society. In both of Maryland's breeding bird atlas projects (1983–1987 and 2002–2006), Jug Bay recorded 120 breeding bird species—the highest number for any location in the state. In all, some 284 species have been observed within the Sanctuary. The western shore of the Patuxent River (in Prince George's County) is protected as part of the Jug Bay Natural Area within the Patuxent River Park and is managed separately, but cooperatively, with the Sanctuary.

Jug Bay itself is a shallow embayment located in the tidal freshwater region of the Patuxent River estuary. The difference in water levels between high and low tides, or the tidal amplitude, is about 30 inches, and salinity varies from zero parts per thousand for most of the year to a maximum of about two parts per thousand in late summer and early fall. The Sanctuary contains about 65 acres of freshwater tidal wetlands in Upper (north) and Lower (south) Glebe Marshes, and a further 199 acres of wetlands belong to other Sanctuary properties upriver. Additional habitats within the Sanctuary consist of scrub-shrub wetlands and swamp, nontidal wetlands, mixed upland hardwood forests, managed meadows, recovering horse pastures, and stream valleys. Much of the forest was logged or farmed during the past 250 years, the current stand being some 50 to 75 years old, with a few trees surpassing 100 years along the slopes of stream valleys and the floodplain. A railroad bed (now abandoned) was built through the wetlands and forest in 1896 and bisects Upper and Lower Glebe Marshes.

The Sanctuary's wetlands are among the best studied in the country. They are used by research scientists from many organizations and disciplines, covering such topics as microbial diversity, the monitoring of nitrogen and other nutrient levels, and greenhouse gas production. Studies have also been conducted on habitat fragmentation, seasonal changes in butterfly populations, and bird species' fluctuations in forest habitat (the latter as part of the nationwide Monitoring Avian Productivity and Survivorship Program, MAPS). Such studies are pertinent in addressing the problems faced in managing natural habitats throughout North America, especially regarding adaptation strategies for climate change.

Along with Sanctuary staff and visiting scientists, volunteers from the Friends of Jug Bay have energized students and visitors (more than 40,000 in 2017) as they participate in habitat restoration, conduct water-quality studies, encounter leeches and their first snakes, and deepen their appreciation of turtle tracking, migrating warblers, and the biological diversity of the Sanctuary's habitats—or simply enjoy the scenery.

Native Americans lived along the Patuxent River since at least 6500 BC. An archaeological dig at Pig Point (a mile north of Jug Bay) uncovered the oldest known artifacts in the Mid-Atlantic states, including pottery, arrow and spear points, and remnants of wigwams, fires, and culinary practices. The site was probably a center of trade in the region and possesses one of the best continuous archaeological records on the East Coast.

While Native Americans navigated the Patuxent River in dugout canoes, Europeans explored its reaches in oceangoing vessels. In 1608, Captain John Smith navigated Chesapeake Bay and sailed up the Patuxent River, finding the water to be 40 feet deep, clear, and brimming with fish. His claim that "Heaven and Earth never agreed better to frame a place for Man's habitation" lured many English colonists to America. The river became host to some of the earliest settlements in the colonial era, and it merited the recognition of the English Parliament during the 1600s as a river of great economic and strategic importance.[2] Ships loaded with hogsheads (large casks) of Maryland-grown tobacco set out for England, and warships, steamboats carrying passengers as well as freight, and hunters in shallow boats coursed the Patuxent channel and its creeks in search of fortune.

It was the Patuxent River that George Washington and, later, Presidents Jefferson and Madison crossed to reach Maryland's capital when America was in its infancy as a republic. And it was on the river that the US Navy flotilla bivouacked to oppose invading English troops during the War of 1812.

The Patuxent itself drains 884 square miles of watershed, from the hills of Maryland's piedmont to the coastal plain. Flowing generally southeasterly, it crosses the urbanized Baltimore–Washington, DC, corridor and opens up into a navigable tidal estuary near the colonial seaport of Queen Anne (now part of Patuxent River Park) before emptying into Chesapeake Bay, 43 miles downstream. Roughly half of its length is considered to be tidal. The river supports more than 100 species of fish, including largemouth bass, catfish, chain pickerels, and yellow perch. It also sustains nesting and overwintering bald eagles, many waterbirds, and a large extended habitat for other indigenous wildlife. Among Chesapeake Bay tributaries, the Patuxent ranks seventh in freshwater flow into the bay.

Johns Hopkins University scientists have studied the history of sedimentation of the Patuxent River and, in a 1966 report, concluded that prior to European settlement, the sedimentation rate was only about 0.05 centimeters (cm) per year— less than one thirty-second of an inch. Forests covered most of the watershed and effectively prevented excess soil loss. As settlers occupied the watershed and farmed tobacco in the 1600s, sedimentation rates increased markedly. By the mid-1800s, many trees were removed for lumber and to plant agricultural crops, further accelerating soil loss through water runoff.

Sedimentation reached its peak at Jug Bay at the turn of the twentieth century,

Defense of the Patuxent River

The Patuxent River has known no greater advocate and defender than Bernie Fowler, an early 1970s Calvert County commissioner, who led the way in a lawsuit filed by downriver Charles, Calvert, and St. Mary's Counties against upriver counties. The lawsuit forced the state, the upriver counties, and the US Environmental Protection Agency to enact pollution control measures. As a result, between 1985 and 2005, the river benefited from a 26 percent decrease in nitrogen, a 46 percent decrease in phosphorus, and a 35 percent reduction in sediment—this despite urban areas increasing to cover 31 percent of the watershed by 2002.

Of Chesapeake Bay's major tributaries, the Patuxent River experiences the most harmful phosphorus and nitrogen nutrient overloads, coming from urban runoff. The river's other major contributors—single localized, identifiable point sources (industrial, sewage, etc.) and now declining agricultural areas (24% of the watershed)—contribute less to the nutrient load. Forested areas cover 43 percent of the watershed.

In 2004, Fred Tutman became the Patuxent's first riverkeeper. His role has been to protect and improve the quality of the river's water and watershed and provide river access and education at the Patuxent Riverkeeper facility in Nottingham, Maryland.

in large part because of the construction of the railroad causeway across the then contiguous Upper and Lower Glebe Marshes. Sharp spikes in 1933, 1954, and 1972 coincided with major hurricanes. In 1972, Hurricane Agnes, the largest to hit the area in 200 years, brought huge amounts of sediment into the Patuxent River and Chesapeake Bay.

It was the largely unchecked erosion from the Patuxent's watersheds in the late 1960s and the building spree of the 1970s that contributed the bulk of the Patuxent's and Jug Bay's most damaging sedimentation, siltation, and pollution levels to date. In turn, this led to the nearly complete destruction of a once thriving seafood industry along the brackish portion of the river.

The river's sedimentation rates remain high because of continuing suburban development, now reckoned at 0.5 cm (one-quarter of an inch) per year, some 10 times that of the pre-European settlement era. As a result, the river's water is invariably chocolate brown. In the 1880s, when steamboats traveled up the Patuxent River to Pig Point, just above Jug Bay, the river was probably four to six feet deeper than it is today. Now only small motorboats and canoes can navigate that far up the river.

Over the past 50 years, nationally recognized land conservation efforts have saved tens of thousands of acres from the Baltimore–Washington, DC, corridor's bedroom sprawl. The southern half of the US Army's Fort Meade was added to the Patuxent Wildlife Research Center, which, at 12,300 acres, is the second-largest contiguous public park / refuge within 30 miles of both the nation's capital and Baltimore. About 8,575 acres of public land, centered on the Jug Bay region, form the fifth-largest Baltimore–Washington, DC, conservation area. It is the largest tidewater conservation area, including the Jug Bay Wetlands Sanctuary, the Merkle Wildlife Sanctuary, and the Jug Bay component of Patuxent River Park.

The wetlands of the Jug Bay area and elsewhere are among the most productive of natural environments and support many plants and animals.[3] The current population of the 937-square-mile Patuxent River watershed, however, is anticipated to increase by 200,000 people—to nearly 1.2 million—by 2020, and this, coupled with climate change and economic development, will place the river under continuing threat. Conservation efforts are being undertaken by dedicated scientists and committed volunteers, and one has to hope that such efforts will be sufficient to retain the river's ecological integrity and protect natural assets that should belong to Maryland's future generations.

To see the major features of the Jug Bay Wetlands Sanctuary, see the map following page 130.

THE SEASONS
BY MONTH

*Each month has two names: the common calendar name
and that given by Native American tribes in the Chesapeake region.*

A view of Upper Glebe Marsh in winter. *Colin Rees, with thanks to Debra Gage*

JANUARY
The Cold or Snow Moon

First January is here,

With eyes that keenly glow—

A frost-mailed warrior striding

A shadowy steed of snow.

Edgar Fawcett, "The Masque of Months"

7 January | THE FIRST SNOW

Snow. The first of the year. It dusts a brooding landscape with a bone-chilling blast from the northwest as winter asserts its icy grip. The marsh echoes with the honking of Canada geese patrolling a meandering creek, some tucking their heads beneath their ruffled wings. A disturbance alerts them to danger and they paddle furiously to rise and circle high over the marsh and neighboring forest, their sonorous calls drowning the strident song of a nearby Carolina wren. A white-breasted nuthatch searches the bark of an oak tree, and a roaming flock of American robins invades the forest, noisily flipping leaves that carpet its floor. In their wake, white-throated sparrows and dark-eyed juncos furiously scratch the exposed areas. Within minutes, all members of this "Gang of Four" disappear. Only the cold wind, rattling tired leaves attached to a stand of beeches, and the distant drumming of a woodpecker are left to the world and me.

All continues subdued and still as I venture along one of the forest trails, the gaunt skeletons of trees revealing the crooked limbs of black cherries, the dark trunks of red maples, and the mighty branches of northern red oaks. Herbs are withered and brown, their seeds dispersed over the snow's crust. The leafy nests of eastern gray squirrels, built high in the skeletal oak trees, are neatly topped with snow and help relieve the monochromatic tone of this midwinter landscape.

I step off the trail to examine the small, creeping stems and leaves of partridge-berries—a mat-forming, evergreen perennial—pushing back leaves on the forest floor. It bears scarlet fruit that matured during the previous fall, doubtless to be consumed with relish by several bird species, as well as by foxes, mice, and skunks as the winter deepens. The scientific name of its genus (*Mitchella*) honors John Mitchell (1711–1768),[1] a Virginia physician who corresponded with famed Swedish botanist Carl Linnaeus. Looking up, I see that the velvet buds of some ash trees are fully formed, awaiting the promise of spring.

Farther on, the smell of the forest is suddenly usurped by a musky odor wafting across the trail and, in time, I spot a red fox, sleek in its magnificent russet coat, loping through the undergrowth. It sits awhile, with ears fully alert. Then, tired of my presence, it quickly continues its lonely journey. Astride a branch, a juvenile Cooper's hawk, elegant in posture, turns its magnificent head and eyes a passing flock of Carolina chickadees and tufted titmice. The hawk takes leisurely wing and passes from view. Against the snow, a northern cardinal male shocks the senses with its resplendent plumage.

I return in late evening to witness the golden fire of sunset, emblazing layered clouds, penetrating woody undergrowth, and gently illuminating snowflakes cascading from branches swayed by the wind. Many trees are adorned with swarms of American robins calling "tut-tut-tut" or a sharp "piik-piik"; others perch in shrubs or scurry across the forest floor. Clearly, this is their roost for the night. By sunset, only a few birds offer up their evening calls and announce the coming darkness.

11 January | WARMTH

What a contrast with my last walk. With a near cloudless sky and temperatures set to reach spring-like warmth (the upper 40s to lower 50s), wildlife is resurgent. Squirrels are bouncing across the landscape, and the raucous cries of blue jays, the full-throttle singing of Carolina wrens, and the gentle clucking and incipient song of American robins fill the air. The snow has gone, leaving a sodden forest floor where birds cavort in unfrozen pools.

I use the spotting scope on the Observation Deck, which overlooks Upper Glebe Marsh (why "Glebe," I wonder?), and find two mute swans gracing ice-free water. Farther off, northern pintails, American black ducks, and buffleheads cruise in the open channels of the Patuxent River. Clouds of ring-billed gulls chatter with mallards, and I hear the occasional whistles of other ducks. In the main channel, three male hooded mergansers take simultaneous dives, watched by an intent gull. The gull flies to the spot of an emergent bird, but indifference reigns and both swim apart. Near the shoreline, adult and juvenile bald eagles perch uneasily on a branch and attempt to peck through the remaining ice. Eventually, the adult wends its way downriver, only to circle back to land in a nearby tree, with its white head for all the world to see.

The marsh is weather beaten, and the low tide reveals an empty mudflat strewn with dead and tangled spatterdocks. Green arrow arums have long been swept away. Dead beds of wild rice and cattails sway in the southerly wind, and only the starkly red branches of gray dogwood on the marsh edge lend color, save for the blue sky.

I walk along the old Railroad Bed Trail, crunching old stony coal residue laced with mud, and encounter the tracks of an eastern gray squirrel, the footprints of a fox, and the hoofmarks of white-tailed deer. In the remaining snow, a mourning dove has left the light imprints of its tiny claws, which trace a delicate path through some scrub.

By chance, I force foraging eastern towhees, white-throated sparrows, and Carolina chickadees to keep ahead of my hesitant advance. A pair of eastern gray squirrels dash up an oak trunk, their moves on land reminding me of rabbits (except that tracks from the front feet of squirrels are usually paired). They stop suddenly and mate, and I imagine offspring will be born some two months hence, with the female becoming fiercely protective of her young and keeping the male distant.

At the Observation Blind overlooking the downriver Lower Glebe Marsh, I ponder swirling patterns of ice, shaped by wind, and the remnants of cattails, with occasional upsurges being forced by the rising and falling tides. In open spots, mummichogs (a type of killifish) disturb the glittering surface, whose sunlit reflections play on the walls of the blind. The name "mummichogs," derived from a local Native American term meaning "going in crowds," is very appropriate, for the school before me numbers hundreds of individuals. In colder times they may burrow in the bottom silt or retreat to deeper waters, and some have been observed to swim under the ice of small tributaries. Spawning occurs from April to August.

Despite the beguiling sunshine, the empty osprey breeding platforms and tree swallow nest boxes (attached to a post supporting the Observation Blind's deck) are reminders that spring is a distant prospect.

Along the Otter Point Trail, dividing forest from Lower Glebe Marsh, the green patches of clubmoss hearten the eye. The stems superficially resemble small seedlings of coniferous trees, and European settlers ripped them up for use as decorative ropes at Christmas. As these mosses are slow growing and thus faced extinction, botanists in the 1990s waged effective campaigns to discourage the practice.

In the autumn, as the leaf blanket settles, it smothers the clubmoss plants. They escape by "running"; hence one of their common names is "running pines." As they spread, they creep from under the leaf blanket, raising their growing tips at the edge. Periodically, the plant sends a branch vertically upward while sinking a root. Though falling deciduous leaves may eventually cover prostrate parts of the plant, loose leaves remain imbalanced on the upright parts, and the slightest wind knocks them off.

At the border of the trail, common chickweeds poke through the leaves, and I have to wonder when colonial boats docking along the river first spread this alien (nonnative) species. It may also have been deliberately cultivated for its curative properties.

My walk ends with the simple tapping of a woodpecker on dead wood. How very powerful it is, and I wonder why I feel it to be a gateway to spring.

12 *January* | VOLATILE WEATHER

Spring-like warmth is delightful, a return to winter this weekend is just spiteful. Temperatures reached 76°F this day.

Washington Post

17 *January* | FISH MOVEMENTS

Fishermen report that yellow perch are moving up into the tributaries of Chesapeake Bay, including the Patuxent River, with white perch right behind them. Spawning is happening, with small males of both species being the first to arrive at their breeding grounds. Chain pickerels and white crappies are soon to follow, the latter schooling and becoming active in fresher water.

Sunrises and Sunsets

> There is then no limit to the multitude, and no check to the intensity, of the hues assumed. The whole sky from the zenith to the horizon becomes one molten mantling sea of color and fire; every black bar turns into massy gold, every ripple and wave into unsullied shadowless crimson, and purple, and scarlet, and colors for which there are no words in language, and no ideas in the mind.
>
> John Ruskin essay, "The Splendors of Sunset"

Autumn and winter sunrises and sunsets are a phenomenon apart. The skies glow with ever-changing hues of pink, red, yellow, and orange and provide radiance to an otherwise gray landscape. But why are they more vivid now than at any other time of the year?

During sunrise and sunset, light passes through much more of the Earth's atmosphere and, thus, a greater number of molecules in the air. Blue light is scattered more easily because its short wavelength allows greater passage for longer-wavelength colors. As Stephen Corfidi, with the National Oceanic and Atmospheric Administration (NOAA), states, "Because air circulation is more sluggish during the summer, and because the photochemical reactions which result in the formation of smog and haze proceed most rapidly at that time of the year, late fall and winter are the most favored times for sunrise and sunset viewing over most of the United States."

So now, in January, weather patterns allow dry, clean Canadian air to sweep across the country, and more colors of the spectrum make it through to our eyes.

19 January | Bird Survey, Moss Carpets, and Beautiful Winter Mushrooms

The pink streaks of dawn give way to golden light as I make my way to participate in the Winter Waterbird Survey, conducted biweekly to track the diversity and abundance of geese, swans, ducks, shorebirds, raptors, and other birds frequenting the wetlands of the Jug Bay area. On this occasion, 15 intrepid birders, all well padded (with clothing) and armed with binoculars and spotting scopes, head for the Observation Deck overlooking the main marsh and its tidal channels. Immediately visible are hundreds of Canada geese scattered along the river. A fair number of mallards take off and land near parties of American black ducks, paddling in distant channels penetrating the marsh. A blizzard of over 1,000 ring-billed gulls (mostly coming from a nearby landfill) noisily rises above the distant cattails, and we see a bald eagle dive into their midst in search of a meal. Later, a northern harrier patrols the marsh, riding and falling on tiny thermals generated by minute shifts in air currents from water to reed. Suddenly it does a flip and plunges into the vegetation to seize its prey. All the while, newly arrived red-winged blackbirds are declaring their territory.

From 2003 to 2011, survey participants counted 108,000 birds, compared with a previous count of 150,000 from 1990 to 1999. Over 83 species have been documented, and the data have been shared on the Cornell Lab of Ornithology's eBird website and with the Google group Maryland & DC Birding. The data reveal high midwinter numbers of ring-billed and herring gulls and waterfowl (Canada geese, American black ducks, and green-winged teal). Flock sizes decline during January and into February, when bitter weather turns the river and its creeks into a frozen landscape for many weeks, forcing birds to seek open water to the south or east, where higher salinities inhibit the formation of ice.

A comparison of the survey periods finds that the number of species (species richness) remains highest in the fall and spring and lowest in midwinter. Although 90 percent of the species have been stable across both periods, peak richness shifted from October and April to November and March and may reflect changing migration patterns.

I leave the birders and walk along Otter Point Trail. Vibrant carpets of mosses and lichens line either side of a small path. On a downed hardwood trunk are spectacular numbers of polypore mushrooms, with their top surfaces sinuous concentric zones of brown, gray, and dark red, and their shape and multiple colors supposedly similar to those of a wild turkey; indeed, they are commonly called turkey tails. A neighboring rotting log is covered by thin-walled maze polypores, or blushing bracket mushrooms. Corky, semicircular in shape, with a concentrically zoned brownish upper surface, their whitish undersides of maze-like pores are quite striking. The openness of the forest allows me to see that other fungi have malformed portions of the trunks and branches of trees, attacking the bark and the inner, or cambial, layers. Such fungal cankers assume a bulbous form on oaks and create knots on cherry trees.

As "ecological engineers," North American beavers can transform a small stream into sizable ponds and so heighten biodiversity, as well as bring other environmental benefits. This is the case at Two Run Branch, where I now stand. A pond appeared in the late 1980s, when a small dam took shape across it, about 100 yards upstream of Otter Point. A few years later, another dam appeared upstream from the first, creating two ponds. Both dams cut off the tidal reach, so the ponds no longer drained at low tide. The permanent water soon began killing trees, with standing dead or dying trees (snags) becoming home to woodpeckers and other hole-nesting songbirds. When dead branches and toppled trunks fell into the ponds, they became basking sites for wood ducks, common watersnakes, and five species of turtles. Increased plant life became rich and diverse; frogs serenaded from the grassy margins of the ponds; muskrats and otters took up residence; red-headed woodpeckers explored tall snags; and fish, aquatic dragonflies, and other insects proliferated.

As Christopher Swarth commented in *Marsh Notes*, North American beavers have a bad reputation, but Smithsonian Institution scientists contend that the spread of beaver ponds across the landscape can have an overall positive impact on ecosystems. At a large-scale level, the landscape becomes a mosaic of current and former beaver sites along riparian networks. Stream corridors change from narrow, defined zones to wider, physically and biologically more complex and productive regions.

22 *January* | Skunk Cabbage Stomp

In winter's false sunshine, a party led by Siobhan Percey, a volunteer, peruses the soaked forest floor in search of a harbinger of spring. The annual "swamp stomp" for skunk cabbages, often described as a "quirky, cunning, and sometimes malodorous plant," has begun.

The flowers of skunk cabbages, a perennial plant, break through earth and snow with a mottled maroon, hood-like bract, or spathe (a modified or specialized leaf), surrounding a knob-like structure called a spadix (a fleshy dark red spike of many petal-less flowers). Because this plant has to attract pollinators in midwinter, its spadix has the ability to internally produce temperatures of up to 60°F–70°F, making it different from other plants and placing skunk cabbages among a small group of plants exhibiting thermogenesis (the process of generating heat). A study by Katherine Thorington suggests that the heat produced by skunk cabbages, beyond allowing the plants to grow in icy conditions, may help spread their trademark fetid odor. Carrion-feeding insects attracted by the smell are encouraged to enter the spathe and thereby pollinate the flowers. By rotating the sex of its flowers, with the spadix initially taking on a female form, the plants avoid self-pollination and thus maintain their genetic diversity.

Following successful pollination, by late spring a tight roll of fresh green leaves beside the spathe unfolds to form huge, dark green, cabbage-like leaves that carpet the saturated soils of moist, shady ravines, to better gather light before the tree canopy covers and shades the ground. As skunk cabbages mature, their contractile roots pull the stems deeper into the soil, so that the plants grow downward, not upward. Each year skunk cabbages grow deeper into the earth, and older ones are practically impossible to dig up. They reproduce by hard, pea-sized seeds that fall in the mud and are carried away by animals or floods. The leaves rot away by August, and the roots are all that remain until nature's cycle begins again.

In the 31 October 1857 entry in his *Journals*, Henry David Thoreau found cheerfulness and hopefulness coming from skunk cabbages: "If you are afflicted with melancholy at this season, go to the swamp and see the brave spears of skunk-cabbage buds already advanced toward a new year. See those green cabbage buds lifting the dry leaves in that watery and muddy place. There is no can't nor cant to them. They see over the brow of winter's hill. They see another summer ahead."

25 January | LICHENS

In melting snow and bright winter light, lichens are more colorful than at any other time of the year. Even on overcast days, they bring a richness of colors—red, yellow, apple green, orange, brown, black, and sometimes blue. Some are like chameleons, on a wet day changing their colors from an earthy brown to bright green.

Both sides of a narrow path near the trail I now tread are bordered by this display. Beyond their color, lichens also bring a richness of sizes and shapes. Round patches of bright green shield lichens, whose shape resembles leaves, are noticeable on most trees or on rotting wood. There are also clumps of powdery goldspeck, a

Gray reindeer lichen. *Jim Brighton*

yellow crumbly looking mass on some trunks, and candleflame, a greenish–lemon-yellow growth, close to the base of an old oak. But it's seastorm lichens that intrigue me. Attached to the reddish bark of some trees, the greenish-gray uplifted edges of its lobes are reminiscent of foam on ocean waves. On the open sandy soil at the end of the Otter Point Trail, I find ground-dwelling, moss-like gray reindeer lichens. Shaped like miniature golf tees, albeit somewhat battered ones, the sprinkling of light gray-green dust (squamules) between them is part of the lichen.

A broad array of surface features and structures help classify and identify lichens: foliose (leaf-like), crustose (thin and crusty), fruticose (shrub-like), squamulose (scaly), and umbilicate (attached to a surface by a single frond). Many different lichens grow on trees and change either through natural succession or fluctuations in their microclimate. Some take up to 30 years to mature.

Lichens play an important role in many ecological systems. They succeed in highly variable environmental conditions by having evolved many adaptations and can endure lengthy periods of dehydration. They break down rocky minerals, leading to soil formation. Birds, insects, and other animals use lichens as food, nesting material, and camouflage. Observations that high levels of nitrogen and sulfur in the atmosphere have a negative impact on lichens have lead scientists, including those with the US Forest Service, to use certain lichen species as indicators of air quality.

26 January | WHERE'S WINTER? PART I

The media are asking, where is winter? With a snow total of 0.4 inches, the region is seven inches below average. With temperatures challenging the norm, December was two degrees warmer than average, and January, a surprising six degrees. There have also been fewer winter storms. A colder pattern lies ahead, with February, the snowiest month, at some 5.7 inches.

30 January | ANOTHER DUSTING AND WINTER FLOCKS

Breezy, cold, with some flurries zipping through and skies clearing from time to time. Not an abomination but less than fair.

Local weather forecast

Memories of the snowy days of childhood renew our sense of wonder, and this morning's whitened landscape is no exception, beautifully reflecting white-blue light into an empty sky. Though we may feel such pleasure, most wildlife species are put at risk and use survival strategies to meet fluffy, dry, wet, crusted, glazed, packed, or drifting snow conditions. Slender branches bend with the weight of snow and ice, and some tree trunks flex, only to spring back as the snow melts. Conifers, intercepting falling snow, prevent accumulations from reaching the ground and provide respite for white-tailed deer and other animals. Some bird species recall hidden stores of nuts, and foxes and other mammals rely on subsurface dwellers for food, sometimes diving into the snowpack to find them.

During the coldest winters, omnivores invariably enter dormancy or become hibernators, whereas herbivores maintain vegetative food caches beneath the snow. Meadow voles store up to six pounds of food. They also huddle in communal nests, to benefit from mutual warmth.

Some invertebrates, including spiders, wasps, and beetles, may thrive in snow, being typically active at temperatures as low as 23°F. At subfreezing temperatures, some produce antifreeze agents in their body fluids, others fast during the winter,

Maryland's First Naturalist: 1671–1702

Thanks to the efforts of an Anglican rector at Christ Church, some 25 miles from Jug Bay, we know what plants grew in the lower Patuxent River and can view specimens he collected, housed in the British Museum. Rev. Hugh Jones arrived in Calvert County in 1696 and began collecting specimens over the next few years, being joined by botanists William Vernon and David Krieg. He shipped back pressed plants and seeds for planting in English gardens, as well as small dead animals.

In 1983, many of the pressed colonial flora were briefly displayed in the Old State House in Annapolis, providing botanists and naturalists with an opportunity to view plants that grew in the Jug Bay area some 300 years earlier.

In a letter dated 23 January 1698, Jones described the local land as a low and "very woody lie one continued Forrest, no part clear but what is cleared by the English.... Indeed in a few years we may expect it otherwise, for the tobacco trade destroyes abundance of timber, both for making hogsheads & building of tabacco houses, besides cleareing ground yearly for planting." Today's botanists have established that nearly 10 percent of the vascular species Jones collected were Eurasian plants brought to the New World; the corresponding figure today is about 25 percent and is much higher in weedy clearings.

Given such a substantial clearing of the area, it is not surprising that a number of the species Jones collected, particularly herbaceous woodland wildflowers, can no longer be found in the Jug Bay area. Some are now confined to Western Maryland, and others have been lost from the state. Tree species have fared better. Except for white cedars, all of the trees sampled by Jones currently grow in the Jug Bay area, although not the majestic specimens he would have seen in what remained of the "ancient forests" in his day. The curious omission of loblolly pine and red maple from Jones's specimens seems to imply that these American species—now common in Southern Maryland—spread into the area as a result of cleared and later abandoned land.

expelling freezing-sensitive contents from their digestive tracts, while still others survive in immature stages, as either larvae or pupae.

Often the silence of winter landscapes is suddenly disturbed by distant bird-calls. Within minutes, a flock appears, bringing color, sound, and movement—scaling trunks, flitting among tree branches, and hanging upside down from twigs—only to quickly disappear. The birds may appear to enjoy one another's company, but the difficult search for seeds and berries has to compete with watching for danger. In a flock, there's a better chance that a swooping hawk or other predator will be spotted in time to sound the alarm and make a getaway.

There are also leaders and followers. In treetop flocks, black-capped and Carolina chickadees are usually the leaders. They stick together throughout the cold months, making regular circuits through the forest. Other small birds fall in with them, following their lead—usually red- and white-breasted nuthatches, brown creepers, tufted titmice, golden-crowned kinglets, and downy woodpeckers. They are clued in to the alarm calls of their chickadee leaders, who are able to spot danger quicker than other birds flocking with them. For example, brown creepers, slowly moving up trees with their faces close to the bark, cannot scan their situation as easily as does a chickadee.

These mixed flocks rarely compete for the same food. Woodpeckers, nuthatches, and brown creepers examine the bark of trunks and large limbs, golden-crowned kinglets explore the tips of twigs, while tufted titmice often hang upside down from heavier branches. This is an elegant demonstration of niche separation, where competing species use the environment differently in a way that helps them to coexist.

Such winter flocks often change composition. Some members have their own winter territories, remaining in the flock only while they travel through their particular portion of the forest. A woodpecker, for instance, will follow chickadees until it comes to the edge of its territory and then drop out, while another woodpecker from a neighboring territory will take its place.

To test the hypothesis that the amount of flocking is related to the availability of food, in 1988 Paul Ehrlich and some students carried out an experiment in two Ohio woodlots. One woodlot was left undisturbed; the other was provisioned in early November with an ample supply of sunflower seeds and beef suet. Downy woodpeckers, tufted titmice, Carolina chickadees, brown creepers, and white-breasted nuthatches participated much less frequently in mixed-species flocks in the provisioned woodlot than in the undisturbed (control) woodlot. This result supports the hypothesis that increased feeding efficiency is a major cause of mixed-species flocking.

January is the coldest month of the year, and, having just passed the winter solstice, each day will be spent in two-thirds darkness. Most plants lie dormant, and many animals have elected to hibernate. Only summer migrants, inhabiting warmer lands, will be free of winter's icy grip.

Winter forest, with snow, along the Otter Point Trail. *Kerry Wixted*

FEBRUARY
The Hunger Moon

Rich meanings of the prophet-Spring adorn,

Unseen, this colourless sky of folded showers,

And folded winds; no blossom in the bowers.

A poet's face asleep is this grey morn.

Alice Meynell, "In February"

2 *February* | TREE SURVIVAL STRATEGIES

The one red leaf, the last of its clan,

That dances as often as dance it can,

Hanging so light, and hanging so high,

On the topmost twig that looks up at the sky.

Samuel Taylor Coleridge, "Christabel"

Broad-leafed trees adopt a variety of strategies to survive winter, the most obvious being leaf drop. Leaves add considerable mass, and their wind drag becomes a significant liability for trees during both snow and ice storms. They also add more surface area for cold, dry winds to suck their moisture through evaporation. Leaf loss also lessens the available surface area likely to accumulate snow and ice buildup during winter storms and threaten to break limbs, if not the trees themselves.

In winter's severe cold spells, while bark's main purpose is to protect the tree from freezing and cracking, bark also protects it from the heat of the sun. It is not necessarily freezing sap that causes a tree to crack, but rather the differential heating and cooling of the tree from inside to outside. At night, however, when bark is exposed to the cold, it can cool and contract much more rapidly than the interior of the tree. This can result in splitting the protective bark layer, due to the circumference of this outer layer of the tree becoming smaller than the warmer, still expanded interior circumference.

There are several ways in which bark protects a tree against this danger. By having a reflective light gray (American beech) or white (river birch) coloring, bark warms up less during daylight hours, thereby reducing the temperature difference between the interior and exterior layers of the tree. Though dark-colored trees will warm up faster than the white-colored trees, their deeply furrowed, or scaled, bark also acts as a radiator to diffuse heat before it can warm the interior layers. Such bark also contains many thin layers that can expand or contract with heat, cold, and moisture. When bark absorbs moisture, it lessens impacts and helps minimize damage from any nearby falling trees and their limbs during rain, snow, and ice storms.

A final adaption: in winter, most of the tree's sap is stored deep underground, in its root system. The small amount of sap that remains in the aboveground portion of a tree has a higher concentration of sugar, thus lowering the freezing point of the sap's water content.

Turning to animals, I am reminded that the breeding cycle of great horned owls is often tied to the weakening of their prey from the cold, a shortage of food, a lack of water, and a decline in vegetative cover. These and other birds require the greatest amount of food when they are feeding their young at the nest—which is now getting underway for these owls.

4 *February* | WHISTLING WINGS

I first hear the whistling wings of tundra swans and then see 11 of these stellar birds with straightened necks approaching and landing in the lead gray water of the main channel of the Patuxent River. Their unique sound led Meriwether Lewis to call them "whistling swans," a name still recalled by the older generation. These elegant, ceramic white creatures now sit midstream, some tucking their necks under their wings, and gently bob in the water.

I muse on visions past when, far up, I see a V-shaped flock migrate northward across the azure sky, their glistening plumage catching the rising sun. Soon the tundra swans will be gone, traveling over 4,000 miles through Midwestern states, Southern Ontario, and the Canadian plains states, arriving at their breeding grounds in northern Alaska and northern Canada between mid-May and

Two magnificent tundra swans. *Frode Jacobsen*

mid-June. The birds then fly south between the end of August and the end of September. Many a lake, swamp, or favored spot to eat and rest has received its name from them, a goodly number in Maryland. Destruction of many of the tundra swans' wintering wetlands has reduced their food sources, but they have adapted by feeding on vegetation left in fields after the fall harvest. These swans are strong flyers, and later the 11 take flight by running across the water and noisily slapping their wings against the surface.

At home, I look at John James Audubon's painting of a tundra swan (a whistling swan to him) and marvel at its pose and setting. I am intrigued by the presence of three yellow waterlilies in the foreground and learn they were considered a figment of Audubon's imagination until 1876, when they were rediscovered and named *Nymphea mexicana*. Yellow waterlilies are now known as a noxious weed outside their native range.

6 February | WHERE'S WINTER? PART II

"Unseasonable" and "spring-like"—the descriptions in the media scramble to reflect the continuing balmy conditions and allocate "the blame" for this change. Angela Fritz, in the *Washington Post*, quotes Jason Furtado of the University of Oklahoma as stating: "A large contribution is coming from the North Pacific jet [stream]. We've been seeing a lot of episodes where the North Pacific jet is strong and extends across the Pacific, and that helps to pump moisture into the West Coast and warmth into North America."

Furtado cautions that while there have been cold snaps on the East Coast, they haven't been sustained. That may have to do with what's going on to our east. "We're not seeing blocking in the North Atlantic," he noted. "That's critical for East Coast winter storms, which almost always require cold air to be blocked and held in place over the Northeast by strong high pressure over Greenland. If there is no cold air dammed up and ready to go, the Mid-Atlantic is simply not going to see that much snow."

February, it seems, is sent to tease—giving and taking, sometimes darker than December, sometimes light as a day in May, and offering both hardness and softness ahead of brushstrokes of green.

10 February | COLOR AND TONE

Describing the color of a crow or swan as black or white is standard practice, but, as artists attest, a closer look will find these terms entirely inadequate to describe their true colors. Black is not the color of a crow, especially if the sun is shining on it. Even on a gray winter's day, its blackness has a different quality than the silhouettes of the surrounding bare branches. The white plumage of an adult tundra swan undergoes many changes with variable light and surroundings and proves to be anything but white. As internationally renowned artist Charles Tunnicliffe states in *Bird Portraiture*:

> Notice the yellow tinge in the feathers of neck and upper breast, and the cold bluish purity of the back, wings, and tail. Note also the color of the shadowed undersurfaces, and how it is influenced by the colour of the ground on which the bird is standing; if he standing on green grass, then the underparts reflect a greenish color, whereas if

he were on dry, golden sand, the reflected color would be of a distinctly warm tint; or again, if he were flying over water, his breast, belly and underwings would take on a colder tint, especially if the water were reflecting a blue or grey sky.

In winter, after a fall of snow: "Now you see how yellow his neck is and, to a lesser extent the rest of his plumage. Note also the reflected snow-light on his under-sides, which makes them look almost the same tone as, or even lighter than, his top surfaces." He adds: "A bird in reality has no local colour, but is always influenced by the colour of its surroundings. To note and depict some of these is a labour for yourself alone."

16 February | HARBINGERS OF SPRING

Because of their distinctive leaves, cranefly orchids, or crippled craneflies, are very conspicuous against the forest floor. They are dull to shiny green above (sometimes with dark purple spots) and a striking purple below. The plant emerges as a single oval-shaped leaf in autumn, overwinters, and disappears in the late spring to early summer, although yielding blooms in mid-July to late August. The blossoms look like a group of craneflies hovering around the stem, and the first part of its scientific name (*Tipularia*) comes from the name of the cranefly genus, *Tipula*.

Spotted wintergreen, or striped prince's pine, is a miniature evergreen herb with whorled leaves arising from its distinct white veins, contrasting with the dark green of the leaf. It grows in patches in the sandy soil of Jug Bay's woodlands, and its white, waxy flowers appear on tall stalks in late July to early August. Since the flowers nod downward, they must be viewed from below to appreciate their beauty.

The common name for a related wintergreen species, pipsissewa, comes from the Muscogee (Creek) Indians' word *pipsisikweu* ("breaks into small pieces"), after its supposed ability to break down gallstones and kidney stones. Native Americans used to make a tea from its leaves to treat rheumatism, urinary complaints, and stomach problems, and its crushed leaves were applied to treat sores and wounds. It continues to be used as a flavoring in candy and root beer.

In the evening, I hear the barking rasp of a male red fox. It's mating time, so I am safe in assuming that this is a warning to nearby males not to enter his territory. I have yet to hear a vixen make her screaming sound to lure males.

Spotted wintergreen, or striped prince's pine. *Nancy Martin*

18 February | WINTER TREE IDENTIFICATION AND ANTHOCYANIN

I join a workshop on winter twig identification, held at the Sanctuary's Wetland Center and given by Earl "Bud" Reaves Jr., an experienced forester in the Maryland Department of Natural Resources. After taking us through the key anatomical features and terminology (refiring my dormant botanical synapses), 11 of us hike through the woodland, putting our honed skills to the test. As we try to identify candidates put before us, we note the advanced state of buds driven by the galloping spring-like conditions, especially Canadian serviceberry (shadbush) and black cherry trees. We stop at a red maple, examining its small, knotty, brilliant scarlet buds and its truly overall red appearance. We learn that in the 1700s, this species was confined to swamps and floodplains, with subsequent fire suppression having encouraged its spread into drier habitats. Reaves points to the very elongated and pointed buds of beech trees; the large, mustard-yellow, egg-shaped terminal buds of mockernut hickories; and cluster buds near the branch tips of oaks. Then come the shiny, light green to brown buds of American sweetgums, ending in a sharp, tapered point, and the yellow-green spherical buds of northern spicebushes.

Beech Trees: Leaf Retention

American beech trees seem to represent a middle ground between evergreens and deciduous trees. Many of their leaves don't fall when they die. Botanists call this retention "marcescence" and wonder what, if any, evolutionary advantage is gained.

While physiologists agree that marcescence is a trait of juvenile plants, there is considerable debate about why some species would seem to be deciduous in all other respects, yet delay the physiological process of leaf shedding.

Some ecologists suggest that it has an adaptive significance for trees growing on dry, infertile sites, a habitat where beeches grow well and outcompete other species. Retaining their leaves until spring slows the decomposition of the leaves (they rot faster on the ground), and dropping them in spring delivers organic material when it is most needed by the growing parent tree. Even small amounts shed at the right time could shift the competitive advantage toward these species on poor sites.

Others suggest that retained leaves, particularly on young trees and on the lower branches on bigger trees, is an effective means of trapping snow, leading to more moisture at the base of the trees come spring. Still others have hypothesized that persistent leaves might provide some frost protection for buds and new twigs. And at least one study suggested that marcescent foliage could be a deterrent to browsing by white-tailed deer and moose. Buds hidden by clusters of dead leaves do not get eaten and thus live to become new shoots and leaves in the spring.

Marcescence may be helpful to trees living in dry, cold, deer-infested environments. But it may also simply be a sign that beeches are still on their way to becoming fully deciduous from their more evergreen past.

Source: Michael Snyder, Chittenden County (Vermont) forester.

We also pay attention to bark. Heavy fluting captures nutrients in rainwater and helps direct them into the tree's root system. Cankers (diseased areas) attack damaged branches, and a fungus (*Nectria*) on a red maple tree comes into view. "Smooth patch disease" is seen on white oaks, apparently caused by oak parchment, a saprophytic fungus obtaining its nourishment by colonizing and decomposing the corky, dead outer layers on living host trees. This results in smooth patches on the normally rough and furrowed bark. No twigs are needed to identify this tree.

We then take the boardwalk to the marsh, edging our way past the brightening boughs of eastern cottonwoods (announcing the rising sap), highbush blueberries, and pumpkin ashes. The latter, also called red ashes, are a large tree of swamps and bottomlands, often noted for their swollen, pumpkin-shaped butt. Red coloration seems to dominate, with the bright red stems of thicket-forming silky dogwood crowding the marsh edge, bordered by smooth alder thickets on the drier, landward side. The thickets lend a year-round color, but redden vividly in early spring, when their branches are openly exposed to sunlight. The degree of redness (caused by anthocyanin pigments in the bark) is determined by light intensity.[1] In the 10 January 1859 entry in his *Journals*, Thoreau observed that the smooth alder "seems to dread the winter less than other plants.... With those dangling clusters of red catkins which it switches in the face of winter, it brags for all vegetation." These trees and shrubs will play host to nesting birds, including Acadian flycatchers, gray catbirds, and yellow warblers. I see a tattered American goldfinch nest from last year.

Research by Panagiota Karageorgou and Yiannis Manetas suggests that red-colored buds and leaves are eaten less by insects than those of young green leaves. It is hypothesized that anthocyanin accumulation masks the strong green reflection that acts as an optical cue for consumers or increases the risk of predator recognition of such insects. The high phenolic (carbolic acid) content of red leaves may also serve to discourage incidental attacks.

22 *February* | SPRING?

It's 70°F and male wood frogs are croaking, having left their winter hibernation to each look for a mate. They mate in just one or two days, with dozens of pairs moving or floating around quiet pools. During a bad winter, the frogs freeze, thawing out when the temperature rises. For the past few months, it's been very dry, so the pools they seek will be in short supply.

Pollen levels are well above normal in the region, due to the abnormally mild weather. Given the expectation of temperatures in the 50s and 60s over the next few weeks, these levels are likely to remain high. According to the media, tree pollen counts have surged ahead of the normal March amounts, and grass pollen has been detected. People suffering from pollen allergies are really struggling.

A mourning cloak. *Frode Jacobsen*

To my surprise, I spot a large butterfly fluttering about in the woods, sometimes resting with expanded wings in the sunniest spots. It is a mourning cloak, with pale yellow dorsal (back) edges to its dark maroon wings. Iridescent blue dots line the black demarcation between the maroon and the yellow. Entomologist Hugh Newman likened the butterfly's pattern to a girl who, disliking having to be in mourning, defiantly let a few inches of a bright underdress show below her somber mourning dress.

Mourning cloaks invariably are the first butterflies to emerge in late winter. The one I see moving between the trees was hatched in the autumn and may have spent the winter in a tree cavity or underneath loose tree bark. This is advantageous, allowing the species to start mating immediately in the spring, rather than having to migrate back to a breeding area prior to mating, along with many other butterflies. Mourning cloaks, sometimes called the Methuselah of the butterfly world, live up to 10 or 11 months. Individuals have one brood per year—emerging in late June—that feeds before estivation (summer hibernation) in July and August. They then awaken and feed through the fall before overwintering.

One may ask, why emerge when there are no plants with nectar? It turns out that mourning cloaks prefer tree sap, which by now has been running for weeks in their habitat. In other cases, the earliest blooming flowers of spring coincide with the first butterflies, while some of the choicest early nectar plants are actually flowering trees, such as eastern redbuds and dogwoods.

23 *February* | Fog and Mist, and Spawning

Were I called on to define, very briefly, the term "Art," I should call it "the reproduction of what the Senses perceive in Nature through the veil of the soul."

Edgar Allan Poe, "Marginalia, part 15"

How right Poe is. Today my eyes turn the desolate February landscape into a work of art. The mist reveals and then veils, surprises but yet confirms. Thus enveloped, I listen and discern what lies beyond such fleeting images.

In the mist-covered cobwebs, lace-like on the cattails, red-winged blackbirds are now singing, having dispersed in early to mid-February to stake out their territories. During this winter, I have seen them in flocks, often joined by common grackles. Other birds are ratcheting up their songs, and I hear their fluttering now, so close by. Light breaks through, making the water luminous and of momentary beauty.

Yellow perch are returning to freshwater tributaries of Chesapeake Bay for spawning. Hundreds make their spawning run from late February to mid-March, feeding on worms, insects, grass shrimp, minnows, and other small fish. The males arrive first, to locate spawning sites, and the females follow. The female releases eggs encased in a long, accordion-like membrane (egg sacs) designed to hang on rocks, brush, or any stream structure that ensures the roe do not settle on the bottom and become covered by silt. Each sac contains 5,000–20,000 eggs.

Yellow perch are threatened by runoff from infrastructure developments that degrade the water quality in many tributary streams of the Patuxent River. Fishing for this species in the tributaries is limited to a minimum size of nine inches in length throughout the year.

Many freshwater fish inhabit the Patuxent River and other tributaries of Chesapeake Bay, with species such as gizzard shad and striped bass beginning their descent into brackish water during the autumn. These and some other anadromous species—which are born in freshwater, migrate to the ocean, and then migrate back to freshwater to spawn—can tolerate brackish waters with salinity as high as 10 parts per thousand. The downstream penetration of brackish water in the bay is greatest in winter, when the salinity level increases with decreasing temperature.

Deft Selection by Squirrels

Eastern gray squirrels select acorns according to these items' suitability for storage. They bury red oak acorns as winter caches but eat those of white oaks immediately. White oak acorns germinate in the fall, and those of northern and southern red oaks later. Red oak acorns send a thickened taproot deep into the ground before the onset of winter and thus bury themselves to escape seed predation. To prevent white oak acorns from germinating, gray squirrels kill them by excising the seed embryo before caching them. Mature squirrels practice this technique more often than juveniles.

Sources: Fox 1982; Müller-Schwarze 2009.

28 February | Bounding Squirrels

 Gray squirrels are a constant accompaniment, and I take them very much for granted. But on this day, I choose to give them equal time.

I see one scampering over the leaves on the woodland floor, flicking its tail as it climbs a tree and peeps around its trunk. It returns to the ground, pauses and sniffs, and then, without warning, burrows beneath the leaves.[2] It emerges with a seed in its teeth and dashes off to devour it out of sight.

We know the eastern gray squirrel, to give it its full name, by its gray fur, but a closer look reveals a brownish tinge and a white underside. Black-colored individuals are not uncommon and exhibit a higher tolerance for cold than their gray counterparts.

At this time of the year, a squirrel has consumed the food it had hoarded in small caches for later recovery. There is great variation in their creation: some are temporary, with the contents to be quickly retrieved; others are more permanent, with stored food items often not fetched until months later. Either way, the retrieval rate is very high, with the squirrels using accurate spatial memory, aided by landmarks and smell, to uncover the caches.[3]

These squirrels sometimes adopt deceptive behavior to keep other animals from retrieving the cached food, pretending to bury food while concealing it in their mouths. They also resort to hiding behind vegetation while burying their food. For some field workers, such behavior implies adherence to the theory of mind thinking.[4]

Looking up to the tree canopy, I see a number of squirrel nests (dreys) of leaves and twigs in the forks of high branches now being repaired to house litters, normally of four individuals.

Often in February, we see the first hints of spring in cloudless sunny days, only to be sent reeling back under snow, ice, and bitter northerly winds. Many of the wildlife species remain hidden, venturing forth only to find food to tide them over. Breeding stirs the activities of some, with song lending a cheerful aspect to gloomy days. Plants slowly respond to the lengthening daylight.

A view of Mark's Pond, a vernal pool. *Rob McEachern*

MARCH

The Wakening or Crow Moon

The air is like a butterfly
With frail blue wings.
The happy earth looks at the sky
And sings.

Joyce Kilmer, "Easter"

2 March | OSPREYS, WOOD DUCKS, TREE SWALLOWS, AND VERNAL POOLS

On my way to Jug Bay, I admire the fiery sun igniting the emerging foliage. Blossoms are beginning to dot the countryside and trees are assuming a green mantle. High winds in areas of lower atmospheric pressure continue from an overnight storm. At Jug Bay, they must be nearing 45 mph, greatly rattling the leaves still attached to American beech trees.

Participants in the biweekly bird count gather on the Observation Deck and soon spot an osprey sitting on one of the marsh's nesting platforms. This is the first arrival of the season for most and is greeted with great excitement.[1] As if to underline the forward march of spring, four wood ducks fly into view, their broad tails and wings allowing them to deftly maneuver as they cross the marsh. Farther down the river, a swarm of tree swallows fills the sky. But the presence of 100 green-winged teal, scattered American black ducks, and dark-eyed juncos exploring the woodland floor reminds one that winter is far from over.

After participating in the regular winter bird count, I join Chuck Hatcher, a Friends of Jug Bay volunteer, in search of vernal pools —temporary pools of water in shallow depressions providing habitat for distinctive plants and animals. The latter are known as obligate species, and at Jug Bay there are four amphibian species (marbled salamanders, spotted salamanders, wood frogs, and eastern spadefoot toads) and two invertebrates (fairy shrimp and fingernail clams) known to occupy the pools. The pools are covered by shallow water for variable periods from winter to spring, but they are often completely dry for most of the summer and fall. Spring breeding begins as early as February and, as the pools are devoid of fish, they allow the newborns of amphibian and other species to develop safely. Many of these spend the dry season as seeds, eggs, or cysts, and then grow and reproduce when the ponds fill again with water.

Nothing stirs on this occasion. The frogs lie beneath the leaves, and there is no chorusing of males in search of mates. I will return later to witness the season's advance.

3 March | FREEZING CONDITIONS AHEAD

Overnight lows are expected to dip into the mid to upper 20s. The last time it dropped below freezing was on 18 February, and prior to that, the region went some 17 days without falling below 32°F—a new warm streak record. Many plants responded and some began blooming, especially in the atmospheric heat dome of Washington, DC. The hope is that the protective casing surrounding their buds will afford sufficient protection. Certainly, wildlife is being sorely tried.

9 March | MOUNT CALVERT

In winter's sunlight, any person boating on the Patuxent River or looking from the Jug Bay Wetlands Sanctuary will have seen Mount Calvert, a plantation house on the west bank of the Western Branch, where it flows into the Patuxent River. It is the only historic building standing at the site of Charles Town. Built of brick, the house consists of a two-and-one-half-story, side-gabled main block to which a two-part shed-roofed kitchen is attached.

In fields to the west, evidence uncovered by archaeologists reveals that Native American Indians were present from the Archaic period (7500–1000 BC) through the Woodland period (1000 BC–AD 1600). Inhabitants of Piscataway Nation villages farmed the rich resources along the Patuxent River until 1684, when an English colonial town was established at Mount Calvert. It was renamed Charles Town in 1696, when it became the original Prince George's County seat. In 1721, the county seat was moved to Upper Marlboro, and Charles Town gradually disappeared. Until the Civil War, Mount Calvert was a typical Southern Maryland tobacco plantation, based on enslaved labor. The plantation house is all that remains, and it now serves as a museum, featuring an exhibit called "A Confluence of Three Cultures."

Thinking about the earlier times reminds me that the Patuxent River basin is a sedimentary wedge of clays, silts, and gravels eroded from the Appalachian Mountains, which range from 10 million to 135 million years old. During the last period of glaciation, some 12,000–25,000 years ago, the basin was heavily forested with spruce, pine, fir, and birch. Following glacial melt and a rise in the sea level, freshwater streams began to form a proto-Patuxent drainage system. With continued inundation, the mouth of the Patuxent River and its tributaries were drowned and inland swamps arose. Upland vegetation changed to forests of oaks, chestnuts, and hickories. By 6,000 years before the present (BP), the river was saline enough to support oysters and, with continued elevations in both the sea level and salinity, by 3,000 BP, oysters reached their uppermost limits in the river, near present-day Mount Calvert.

The first colonists discovered a vast expanse of seemingly unbroken hardwood forests, dominated by chestnuts, mockernut hickories, river birches, and American sweetgums. In the early seventeenth century, Father Andrew White, a Jesuit missionary who chronicled the early days of the Maryland colony, wrote: "Fine groves of trees appear, not chocked with thorns or undergrowth, but growing at intervals as if planted by the hand of man, so that you can drive a four-horse carriage wherever you choose through the midst of trees."

It was the fertile soils beneath, however, and the deep harbors of the Patuxent River that offered commercial prospects. In time, deforestation and poor agricultural practices left a depauperate landscape, so that by the mid-nineteenth century, over 85 percent of the forest cover in the watershed had been cleared. Sedimentation in some parts of the estuary was five times higher than pre-European rates. Further land clearance brought dramatic increases in runoff.

13–14 March | A MASSIVE SNOWSTORM?

The media are again playing up the chances of a powerful nor'easter roaring up the Mid-Atlantic and Northeastern coasts this coming Monday and Tuesday nights. It is predicted to unleash heavy snow and howling winds. Snow accumulation could be from four to eight inches, depending on differing information from

Pollution of the Patuxent River

A study by Humaira Khan and Grace Brush in 1994 found that sedimentation was particularly evident in the mid-1960s through the mid-1970s and again in the early 1980s, when urbanization was at its highest rate. A subsequent publication by Elaine Friebele and colleagues in 2001 estimated that the annual load of sediment accumulating in the Patuxent River increased from 160,000 tons in 1950 to 710,000 tons in 1980. A growing population, chemical fertilizers, detergents and other contaminants, overfishing, and nonindigenous pathogens brought dramatic declines in water quality and in submerged aquatic vegetation, so that by the mid-1970s, entire sections of the river were declared biologically dead.

Subsequent management efforts have reduced phosphorous and nitrogen levels, and fishing regulations have produced some improvement, but efforts to restore submerged aquatic vegetation have been mixed. A more recent study by the Maryland Department of the Environment in the vicinity of Jug Bay shows that its water quality is fair and that nitrogen, phosphorus, and sediment levels are falling, though they remain relatively high. Habitat quality was judged to be poor, due to low water clarity and moderate abundances of algae. Underwater grasses covered larger areas in the early 2000s, meeting restoration goals, but they have not been as widespread in more recent years and were especially limited in 2012. Bottom-dwelling animal populations were considered to be healthy.

The main channel of the Patuxent River is now 6–10 feet deep, compared with a depth of 40 feet when Captain John Smith explored the river in 1608.

---

the US Global Forecast System and the European models. Meetings are being cancelled and people are on a foraging spree.

By midafternoon, the sky darkened, the wind stilled, and birdsong was at a premium. At 7:00 pm, rain mixed with sleet began covering trees and plants. Later, the wind picked up and pounded trees beginning to be laced with ice. By 5:00 am the following day, many a conifer was bending from the increasing weight, and dawn's light revealed an icy terrain. The snow is barely two inches deep, but a thick crust has formed, especially on evergreens.

I see a few animal tracks and ponder the impact of ice storms on wildlife, for freezing rain brings many a peril. Feathers and fur covered in ice reduce the owner's ability to keep warm. It also adds weight to wings and tails, so extra energy has to be burned, and a slower progression makes birds especially vulnerable to prey. Thick crusts of ice trap small burrowing mammals, sometimes preventing them from reaching the surface to breathe fresh air. This, in turn, shuts down the amount of prey for red foxes, owls, and hawks. Prolonged icing can also deny juncos and other sparrows access to seeds.

Should cold weather persist late into spring, migrant songbirds and some shorebirds could starve from insufficient food, as they would be weakened from their long journey. Early returning American woodcocks are particularly vulnerable and often turn up in suburbia in search of food.

Tree damage by glazing ice can range from loss of tissues to structural failure or fatal injury, with young trees being the most vulnerable, having less resistance than seasoned wood. Even if nonlethal, ice often presents serious physical consequences. Trees having stout trunks, strong wood, symmetrical crowns, limited surfaces on which to accumulate snow and ice, favorable branching, and solid conditions for rooting are the most likely to survive, though site conditions will temper the outcome. The presence of woody vines can exacerbate vulnerability.

On the plus side, ice storms increase the number of habitat-forming snags and the amount of coarse woody debris, initiating stem and branch decay and producing conditions favorable for the development of tree cavities. They may also change the habitat, favoring wildlife species that prefer the early stages of vegetative succession. Conversely, fruit and seed production may be adversely affected if a plant's reproductive structures, developed in the autumn, are destroyed.

A revealing study in central Vermont by Steven Faccio showed that although the total abundance of breeding birds decreased following a 1998 ice storm, species richness and diversity increased. Three species inhabiting forest interiors were less populous following the storm: two canopy foragers (red-eyed vireos and blackburnian warblers) and one ground forager and nester (ovenbirds). Dark-eyed juncos were the only ones whose numbers increased significantly after the storm, although Canada warblers and winter wrens showed increasing trends.[2]

Although midafternoon had seen some relief after a snow flurry, the approaching night brought strong freezing winds, dipping temperatures into the low 20s.

Trees continued to fall, and birds sought the Wetland Center's bird feeders in great numbers. Trees in the budding stage may well suffer damage, and some may not bloom. Various plants are holding back, waiting for the length of day, rather than the temperature, to change. Winter is still with us.

16 March | ICY CONDITIONS CONTINUE

March has produced a surprising dose of winter, following a freakishly balmy February that broke nearly 12,000 local daily records for warmth in the United States. An article by Seth Borenstein notes that last month the average temperature was 41°F—seven degrees warmer than normal, as reported by the National Aeronautics and Space Administration (NASA). Gabriel Vecchi of Princeton University stated that he could not recall such a February and expected this to happen with more frequency in the future.

This morning the marsh is an icy glaze, and bitter winds keep the temperature in the 20s. The usual suspects are in evidence, but there are exceptions. Two immature bald eagles tear apart a fish on the shoreline, while three others fly over the marsh in slow, measured, ponderous wing beats, scaring up the gulls and waterfowl. These juveniles can overtake and master their prey with ease. Two adults soar before us and perform their famous courtship flight, flying up over the marsh, locking talons, and then doing a cartwheel spin as they fall toward the ground, breaking apart at the last second. Though there are no reports of breeding eagles in the Jug Bay area, two parent pairs in Washington, DC, have been shielding their eggs through sleet, snow, and slicing winds, and on 15 March a female nesting at the capital's southwestern police facility was viewed on camera, shifting her weight to accommodate a new eaglet and her other egg.

Before leaving the Observation Deck, the bird count group spots two ospreys perusing the marsh. Doubtless, the birds are surprised by the arctic conditions, as they expect a temperature range of 35°F to 52°F on their traditional arrival date, St. Patrick's Day, which is tomorrow.

On the edge of one of the woodland trails, the party notices an American mud turtle. It is very cold to the touch and shows no signs of life. No doubt enticed out of hibernation by the earlier warm weather, it now suffers the consequences.[3]

As we thread our way through sodden woodland, avoiding the pools, puddles, and rivulets from fast-disappearing snowdrifts, we see one of the first signs of returning life. Our eyes slowly discern a bit of pale blue and, as we close in, we find a little hepatica, with its face shimmering in the partial sun. From the mass of leaves of varying brown colors and shapes, each flower rises on its own stem, covered in long, fine hairs.

On the way back to the Wetlands Center, we have fun spotting animal tracks (mostly of red foxes) in the remaining snow. We reach the building just in time to wish the educational staff success in teaching sixth graders all about the world of amphibians.

18 March | Spring Walk and Vernal Equinox

Every year Mike Quinlan, master naturalist and Sanctuary volunteer, conducts a leisurely hike through the Sanctuary to see and hear the signs of spring and observe the setting sun close to the vernal equinox. This time, there are 17 participants, well wrapped against the northerly wind on a still-sunny afternoon. We are informed that dry and cold conditions have held back spring by some two weeks.

Nonetheless, cranefly orchids, also known as crippled craneflies, seem to fare well. Virginia springbeauties are also now much in evidence, displaying small pinkish-white petals and slender, spear-shaped leaves. Their roots are reminiscent of tiny potatoes, and this species is often given the name "fairy spuds." In his seminal book, *Stalking the Wild Asparagus*, Euell Gibbons extolled their nutritional value and suggested the roots were close to chestnuts in flavor.

We halt before a massive oak tree, to be informed that it is a "witness tree," marking the boundary of an old farm, now occupied by an oak and beech forest. Colonial settlers divided the land, and surveyors documented the trees that rested at the imaginary corners and angles of the parcels to mark their boundaries. What stories this oak could tell! A ditch runs past the tree and probably served as a drain for fields of tobacco, very much the local crop. There are some American beeches that have clearly been pollarded (pruned) in past times to promote a dense head of foliage and branches, the latter used to produce the upright poles favored for fence rails and posts, or for firewood.

The drought of the three previous years continues, and the site of a vernal pool where we now stand is dry. At this time of year, it normally contains up to three feet of water, and the deficit must be having an adverse impact on dependent fauna and flora. We move on to Mark's Pond, created as a borrow pit through the removal

The underside of a cranefly orchid, or crippled cranefly, leaf (*right*),
amid the forest floor (*left*). *Bill Harms*

of soil used to construct the railroad bed in the 1890s. It is now barely filled with a
foot of water. Mike scoops out some wood frog tadpoles and portions of fist-sized
egg masses for close inspection. He also turns over logs and finds an adult spotted
salamander, displaying yellow spots against a black body approximately five inches
long. On rainy nights when the temperature rises, the salamanders make a sudden
migration toward their annual breeding ponds for mating. Adults only stay in the
water for a few days, and then the eggs hatch in one to two months. Our next find
is a smaller marbled salamander, with whitish spots and bars on a purplish-black
body. On rainy nights in late summer and early fall, it migrates to an adjacent area
surrounding a dry vernal pool to lay between 50 and 200 transparent eggs in a
nest under moss, leaves, or other types of cover. The female remains with the eggs
until they hatch, but there is a strategic decision behind siting them. If the eggs are
placed within the pool, they might be washed away with just a light rain. Eggs laid
at too high an elevation might not be flooded by fall rains, but they have to endure
exposure to predation and cooling temperatures. Not surprisingly, most marbled
salamander nests are constructed at midpoint locations.

We walk on, and our attention is drawn to packs of leaves placed in a mesh bag
in a stream whose surface reflects the brilliant scarlet of budding red maple trees.
Every year, schoolchildren and their teachers open up these packs and study the

Value of a Tree

One of the papers on environmental benefits I find most persuasive was published by Tarak Mohan Das of the University of Calcutta in 1979. He quantified the ecosystem services rendered by a tree during its average life span of 50 years. These included producing oxygen and sequestering (storing) atmospheric carbon dioxide; supplying shade and shelter; cooling and cleaning the air; providing habitats; acting as windbreaks; preventing soil erosion; regulating runoff and groundwater; manuring (fertilizing); and aiding nutrient recycling. In 2013, the total value of these services by a healthy tree was calculated at US$550,000.

The most important service trees offer to mankind is the production of oxygen. According to Environment Canada, that country's national environmental agency, a healthy tree, on average, produces nearly 260 pounds of oxygen each year, while a 1988 US Forest Service report indicates that a typical person consumes 386 pounds of oxygen during that period. Thus two medium-sized, healthy trees can supply more than enough oxygen to sustain a single person over the course of a year.

Sources: Das 1979; Helmenstine 2017.

different types of aquatic insects within them to determine the health of the stream and obtain a better understanding of aquatic ecology.

The Beaver Pond, with its two North American beaver lodges (dome-like homes), is silent and still, except for the cries of a passing belted kingfisher and the incessant calls of red-winged blackbirds, American robins, and Carolina wrens. We note the number of loblolly pines downed by the storm of the past few days and the early blossoms of a northern spicebush, with its sweet odor. An American woodcock strolls across our path, and up high, a skein of 100 Canada geese head north in a near perfect V formation.

Clouds obscure the setting sun, but we know we are only two days away from the vernal equinox, marking the beginning of spring and the end of winter.

20 *March* | DEATH OF CHANDLER ROBBINS AT AGE 98

Many revered Chandler "Chan" Robbins as the father of modern ornithology. His *Birds of North America: A Guide to Field Identification*, published in 1966, has sold over six million copies and become a source of untrammelled pleasure for many. Chan worked at the Patuxent Research Refuge for over 60 years, and his contributions were phenomenal. His research, along with that of Derek Ratcliffe, concerning the thinning effect of DDT on the eggshells of ospreys and eagles provided Rachel Carson (a colleague at the time) with arguments in favor of controlling pesticide applications, which she later used in her book *Silent Spring*. He founded the North American Breeding Bird Survey in 1965 and participated in the National Audubon Society's Christmas bird count for more than 80 years. His breeding bird survey happily combined people's passions for birding with driving cars and became the gold standard for understanding avian population trends. Chan was also credited with banding over 150,000 birds, naming as his favorite the house wren, for its "amazingly high-pitched and intricate song."

Chan visited Jug Bay last summer. Though confined to a wheelchair, he was able to observe birds at Plummer House and butterflies at the Butterfly Garden in Glendening Preserve.

In his foreword to the 2010 *Second Atlas of Breeding Birds of Maryland and the District of Columbia*, Chan wrote:

> It is now the responsibility of every citizen to seek ways to further reduce our cumulative impact on avian populations. For example, more Maryland counties should enact cat indoors legislation. US cities should follow the Toronto City Council's lead in mandating bird-friendliness in the construction of high-rise and non-residential low-rise buildings and in reduction of nighttime glare and light trespass. And we must all strive to reduce avian casualties from clear and reflective glass windows, which are estimated to take the lives of a billion birds annually in our country. We must drastically reduce the hazards that we construct along migratory corridors. We should slow down when driving on narrow rural roads in the early morning. Every landscaper and home owner should plant native trees and shrubs that are preferred by our native birds rather than exotic species that are of little use to them. And finally, we must wean ourselves away from the vast biological wastelands called lawns that are now the major crop in

our suburban counties; they provide neither food nor shelter for any of the birds we would like to protect.

These words behoove us to continue Chan's seminal work.

23 *March* | A Swallow of an Evening and a Friends of Jug Bay Meeting

Before I join a board meeting of the Friends of Jug Bay, I walk through a portion of the Sanctuary with Dotty Mumford, a long-standing Jug Bay volunteer and membership chair of the Friends. We stop at the Observation Deck, peer into a northerly wind blowing over a marsh, and watch a retreating tide. A few American black ducks and blue-winged teal gather midriver, and a pair of ospreys, carrying nesting material, circle and land on a breeding platform. On one of the exposed muddy areas, three greater yellowlegs probe for food. Dotty remembers the time when a variety of species of shorebirds came in fairly large numbers. With climate change and rising waters, their foraging area has greatly contracted, and the calls of these intriguing birds are now rarely heard.

We follow the Railroad Bed Trail, with the old coal residue scrunching beneath our feet. The roots of trees lining the sides have penetrated the rotting rail ties, their spacing requiring care by walkers.

A northern rough-winged swallow passes before the setting sun and, at a bird feeder, we notice American goldfinches now sporting their breeding plumage. The skunk cabbage patch at the base of an intermittent creek close to the Wetlands Center building is well advanced, with bright green leaves pushing up in vertical, rolled-up spires a foot high. Many birds eat the emerging leaves, and common yellowthroat warblers have been observed nesting in the central hollows of the large leaves.

The 15 board members of the Friends of Jug Bay meet every two months to decide how to raise and disperse funds, monitor issues that could affect the Sanctuary, and support the staff's education, research, and strategic planning programs. The Friends' backgrounds include professionals (mostly in the sciences) from the federal and state governments and from nonprofit agencies. They, along with volunteers, sponsor lecture series, children's events, canoeing, educational internships, science summer camps, fellowships, and research programs. The latter include bird, herpetology (amphibian), fish, and forest surveys; seasonal changes in the timing of butterfly behavior; water chemistry monitoring; macroinvertebrate sampling; and the impacts of climate change on wetlands.

Last year, volunteers provided 5,300 hours of service, a commitment that amounts to nearly $125,000 of in-kind contributions to the Sanctuary's county, state, and federal partners. For these and other reasons, in 2017 the Maryland Recreation and Park Association nominated the Friends for the association's Outstanding Service Award, which they won.

News flash: Fishermen tell me that the run of yellow perch is nearly completed, and that one by white perch is now starting, the recent low temperatures having slowed the latter's migration. Least brook lampreys are making an appearance in some streams, moving stones with their suction-like cup to create pits in which to deposit their eggs.

25 March | SPRINGTIME SEESAW

While this day provides a taste of spring, with temperatures in the 70s, the transition from winter is never easy. A wedge of cold air sliding down the coastal plain from the north presages cooler conditions tomorrow, but forecasters predict warmer air should begin to surge back over the following few days.

Whatever the outcome, birds are in full throttle: the dulcet strain of American robins, the cooing of mourning doves, and the high, slurred whistles of red-winged blackbirds.

Some people have claimed that with climate change, "real" snow days are going to become less frequent, but more higher-intensity storms and an increasing frequency of irregular weather patterns will occur, as has been witnessed during this month.

According to the USA National Phenology Network's spring indices, spring is arriving 20 days earlier than the long-term average (1981–2010) in much of the South, the Great Plains, the Great Basin, the Mid-Atlantic states, the Midwest, and parts of the Northeast.

Studies of such changes in phenology—the timing of seasonal biological events such as leafing out, flowering, migration, and reproduction—show that already-documented changes in the climate are causing significant modifications to the seasonality of many organisms. For example, blooming earlier can disrupt the critically important link between wildflowers and the arrival of birds, bees, and butterflies that feed on and pollinate the flowers.

Generally, areas that experience extremely early springs are more likely to be susceptible to ecological mismatches and species invasions. Nonnative species tend to have phenologies that are more malleable than those of their native counterparts, enabling the aliens to better adjust to earlier and less reliable spring conditions. Changes such as the phenological responses of rare species will help determine when the monitoring of natural resources in general should occur. This may require parks and other reserves to plan for earlier events and employ more-flexible management schedules and local adaptation approaches to the impacts of climate change.

27 *March* | Peepers: A Wave of Springtime

It's been a glorious spring day, and with temperatures reaching the mid-60s, hundreds of spring peepers are now a-peeping in the dusk. These were first heard a couple of weeks ago, but cold fronts intervened and quieted them. Their sharp calls chorus from the wetland; for some, their cries resemble the sounds of sleigh bells. Each male is calling for females and defending his tiny territory, the single "peep" being his advertisement call. Should a rival male threaten his territory, he will utter a short, trilling, aggressive call.

The "peep" is made by squeezing air over the frog's vocal chords, which is then amplified in the echo chamber of its balloon-like throat sac. A large chorus of singing males may reach 120 decibels and, up close, is truly deafening. Even when calling, the peepers may be difficult to locate, because of their diminutive size (0.7–1.3 inches) and cryptic coloration, or camouflage.

The peepers hide from their many predators during the day, emerging at night to feed. They mate and lay their eggs in water and spend the rest of the year in the forest. In the winter, they hibernate under logs or behind loose bark on trees, waiting for the spring thaw and their chance to sing. They can survive at the point of freezing for up to a week, as their blood possesses a biological antifreeze, preventing immediate death.

During the daytime, peepers often call during light rains or in cloudy weather.

They are usually silent at the end of summer, but resume their calls from forests during the fall.

30 *March* | FLOWERS AND WARBLERS

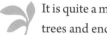 It is quite a morning, beginning with a stunning sunrise caressing the greening trees and ending with more heralds of spring.

From the Observation Deck at Jug Bay, some birders admire the constant flux of tree swallows over the marsh and the mating displays of ospreys. Someone spots a palm warbler bobbing its tail as it moves through the undergrowth along the marsh. This newly arrived migrant, sporting full breeding plumage—a chestnut cap and yellow underparts—is a most welcome sight.

Lyre-leaved rockcresses are increasingly evident along the trail. But it is newly emerged flowers that signal a resurgent spring: Canadian serviceberry bushes with their beautiful, erect, white lacy blooms; the bell-shaped pale pink blooms of highbush blueberries; the yellow showy clusters of northern spicebushes; and the purple-pink of eastern redbuds. All are in contrast with the browns and grays of the landscape. For some, the pure beauty of redbuds encapsulates a moment, described by Emily Dickenson in "A Light Exists in Spring," "that Science cannot overtake / But Human Nature feels."

Later, I hear a short trill reminiscent of another migrant and search for a pine warbler. In the topmost branches of some pine trees (where else?), I find a male methodically prying into needle clusters. He is a somewhat plain creature, with a yellow breast, olive green back, and white wing bars, once described as a watercolor sketch that an artist left unfinished. The "witch-witchy-witchy-witchy" song of another migrant reaches us, and I catch the delight crossing the faces of the observers. It is a common yellowthroat, but not so its plumage. The male has a black Lone Ranger–type mask, with a yellow throat and upper breast. The bird is something of a skulker, and it takes a while to scout the woodland bordering the marsh, where we finally find this wren-like warbler. Common yellowthroats are the most widespread and numerous warblers in Maryland, and they were one of the 10 most frequently recorded birds during both of the statewide breeding birds atlas projects. It is also one of the first bird species to be catalogued from the New World, when Carl Linnaeus described a specimen from Maryland in 1766.

Waterfowl are now departing.

31 *March* | RAIN COMETH

Much-needed heavy rain for the last 18 hours should restore the vernal pools. Some have tadpoles, and whirligig beetles, water boatmen, and water striders are now skating on the water. Dragonfly and fishfly larvae are beginning to stir, the latter having two-inch-long, stoutish bodies and pinching mouthparts used to seize their prey. Fishermen once commonly used fishfly larvae as bait, but polluted streams have made these insects less common.

At this time of year, I am reminded that the seasonal cycle is best appreciated when streaks of sunlight pour through the delicate greens and reds of buds and new leaves to caress the bottom of woodlands. Patches of wonderfully blue sky fill the spaces between dead or bare branches. And even though a cold northerly wind blows, the growing green roof hints at sultry days ahead, when mature canopies will darken the woodland and give shade and coolness.

What a month it has been: cold and then mild weather playing out in the death throes of winter. Now, vernal equinox winds blow through tall trees and send racing waves between brown decayed reeds.

March exemplifies awakening, with longer days and mounting temperatures. For much of the wildlife, it is the dawn of the year. The first blossoms appear; overwintering butterflies and bees emerge, along with amphibians; and loud and vigorous birdsongs fill the air as territories are being established. Winter visitors begin their trek northward, and summer visitors are well on their way to temperate climes.

The boardwalk overlooking Upper Glebe Marsh, with mountain laurel.
Jug Bay Wetlands Sanctuary

APRIL

The Grass Moon

The sun was warm but the wind was chill.

You know how it is with an April day.

Robert Frost, "Two Tramps in Mud Time"

4 April | EVER-CHANGING WEATHER

I have counted 136 kinds of weather inside of 24 hours.

Mark Twain, note on spring

After days of April rain, when birds stopped singing, tree trunks darkened, the woodland floor became peppered with tiny fallen leaves after only days of life, and new grass bent with silver droplets, the landscape dried in a southerly breeze. The meadows within the Sanctuary greened, much like moss on a stone. The wind blurs the green of trees, and American robins have a greener carpet at their feet, their beaks and plumage shining in the midday sun. Songs rise spring-like from both field and woodland.

The weather experts report that March was colder than February for the first time in 33 years. The month of February was the warmest on record, warmer than a normal March. A false spring in February it may be, but the true character of the season will surely triumph.

White-throated sparrows are still around.

5 April | AN AFTERNOON WITH WOOD DUCKS: LOVELIEST OF ALL WATERFOWL

The afternoon presents a cloudless sky, gentle winds, and temperatures in the low 70s, cooled under a noticeably greening woodland. Carpenter bees and honeybees buzz in a fully blooming eastern redbud tree within the Patuxent

Changing Climate over 100 Years

A report on the changing climate of Chesapeake Bay and its tributaries reveals that frost coated the region about 100 times a year in the early 1900s. A century later, there are 30 fewer chilly days. Nights when temperatures do not drop below 68°F have grown to a similar extent, and rainfall has increased to 4.5 inches per week, a gain of 12 percent.

As a consequence, the length of the growing season has expanded to more than 30 days and the impacts on plants and animals are proving to be significant. Flowering trees and plants will bloom earlier and, in general, grow later into the year. Many animals in the bay use changing temperatures as a cue to start such activities as migration, or to begin or end hibernation. If trees bloom too early, a late frost may kill the flowers, eliminating the trees' ability to bear fruit during the year. Some species of submerged aquatic vegetation become stressed and die when water temperatures exceed 68°F for prolonged periods. Equally, the delayed onset of winter may be interfering with the migratory behavior of some species of fish, and warmer winters and springs are increasing the likelihood of producing harmful bacteria. Lastly, increases in precipitation over the past century have augmented the volume of water flowing out of the Chesapeake Bay's rivers, along with nutrient and sediment loads.

As a specific example, if a Baltimore oriole, during its spring migration, arrives too late, it may find that prime nesting areas have been taken up by other birds, thus interfering with the bird's ability to find a mate or successfully raise its young. Equally, some migratory species that call the Chesapeake region home during the spring and summer may have a problem with the delayed onset of fall and winter— missing their cue to migrate and therefore being ill equipped to handle sudden drops in normal wintertime temperatures.

Source: Delgado 2011.

River Park. The tree's unique branching creates a handsome, spreading crown. Eastern tiger swallowtail butterflies flit about the informal garden, and eastern bluebirds have begun brooding their eggs. Summer ducks, more popularly called wood ducks, are appearing—time to put up more nest boxes before their egg laying gets underway.

Marion Clement, the former executive director of the Maryland Bird Conservation Partnership, and I join Greg Kearns, the park's naturalist, to position some nest boxes near the marsh edges along the Patuxent River. Greg has an obvious love for the area and knowledge extending over three decades. He is a world authority on soras and a tireless advocate of bird conservation, mesmerizing audiences with his enthusiasm and good humor.

Relaxing fishermen ring the platform as we launch our boat, laden with new nest boxes of cedar (fashioned by Discovery Channel participants), poles, and equipment. The 50°F water has been warmed by the mild winter, and I observe signs of resurgent marsh vegetation, with spikes of cattails poking through the mud. An osprey lies low on its nest platform and is not to be budged by our passing craft—a sure indication that it has laid eggs. Laughing gulls make their leisurely way upriver and are easily overtaken by the strong, swift flight of a pair of greater yellowlegs.

Greg sinks 10-foot poles (normally used to support roadside signs) into the marsh bottom, and we assist him to line up and secure the nest boxes.[1] The people who made these boxes, which are painted in almost psychedelic colors, obviously had fun, and we wonder if the occupants will appreciate the humor. Some bird-watchers almost certainly will.

Greg opens existing boxes suspected of possessing eggs, and we peer in one to see 19 of them, neatly enfolded in down. This is clearly a case of dump nesting, where two or more hens have deposited eggs, as a normal clutch ranges from 10 to 15 eggs. A clutch of 29 eggs was recorded last year, and it had an 85 percent successful hatch! In dump nesting, the egg-laying interloper slips in when the host hen is absent. Such parasitism is likely where nesting densities are high and nest cavities are in short supply. Another box is full of osprey, Canada goose, and duck feathers and was probably occupied last year by an invading tree swallow.

Wood ducks are stunning. Both sexes have crested heads, and the male is a kaleidoscope of iridescent green, blue, purple, black, and white. His eyes and eyelids are red, throat and breast brown, and sides and belly canvas brown. The female sports a smaller crest, with a white teardrop, and a grayish-brown body.

In pre-colonial times, wood ducks were probably the most abundant species of waterfowl in eastern North America. Unfortunately, overharvesting, coupled with the destruction of wooded swamps, nearly drove them to extinction by the early twentieth century. The Migratory Bird Treaty Act of 1918 is credited with coming to the rescue, along with the use of nest boxes.

The population of this beautiful duck species has greatly increased over the last several years and has become widespread. These waterfowl may well be forced north by climate change, however, putting them under some stress. The National Audubon Society's climate model predicts a 69 percent loss of their current summer range by 2080, as it gradually moves into northcentral Canada.

As I look at one now crossing the marsh, I am reminded of John James Audubon's deft description in his nineteenth-century *Birds of America*: "The flight of this species is remarkable for its speed, and the ease and elegance with which it is performed. The Wood Duck passes through the woods and even amongst the branches of trees, with as much facility as the Passenger Pigeon; and while removing from some secluded haunt to its breeding grounds, at the approach of night, it shoots over the trees like a meteor, scarcely emitting any sound from its wings."

9 April | Submerged Emergents and the Canopies Restored

It is a golden afternoon as I sit with our son Christopher on the Observation Deck. A balmy southerly wind bathes the open marsh and us, and we are content to sit and stare. The unmistakable cries of laughing gulls and the noise of a passing motorboat are all that break the intoxicating silence. There are no waterfowl to be seen. The woodland behind us is sprouting leaves at an almost visible rate, and pollen saturates the air.[2] Full canopies—green cathedrals—seem only days away, but they bear significant consequences.

Different plants growing below tree level in woodlands use seasonal changes in light and temperature throughout the year. They include spring ephemerals, summer greens, winter greens, and evergreens. Each group has a unique survival strategy. Spring ephemerals leaf out in the early spring and die back prior to the closure of the tree canopy. They are shade tolerant, endure low temperatures by remaining close to the ground, and photosynthesize as soon as their leaves expand. Low soil temperatures, which constrain water absorption, heighten these plants' photosynthetic rates.

The majority of understory species in deciduous forests are summer greens, emerging prior to canopy closure, when temperatures have increased. Unlike spring ephemerals, they retain their leaves until they die back in late summer or early fall. Summer greens gain over half of their annual carbon dioxide during this short window of adequate temperature and sunlight.

Winter greens and evergreens retain their leaves during the winter, when spring ephemerals and summer greens lie dormant. Such greens possess leaves produced during the late summer or fall, losing them the following late spring or summer, while evergreens retain their leaves for more than one year. Both gain most of their carbon dioxide during the spring and fall, when temperatures are moderate and light levels highest.

At full tree canopy, insect populations will increase dramatically and gorge on the tender buds and leaves. Migrating birds, in turn, arrive to feed on this insect feast and build their nests behind the foliage without too much fear of predators. The canopy will militate against rising air temperatures, though stream and river levels will drop as trees absorb water needed for their growth.

We walk along the Railroad Bed Trail to check on the marsh vegetation, brown when seen from the Observation Blind but an impressive green close up. Broad- and narrowleaf cattails are pushing through the mud, in some instances reaching 12–15 inches in height and beginning to sway in the wind. Types of vegetation known as emergent species, which grow in shallow water from perennial bulbs or underground plant stems (rhizomes), are beginning their march toward full expression. They include spatterdocks (yellow pond-lilies), with their heart-shaped leaves now rigid at low tide; the unfolding bright green, arrow-shaped leaves of green arrow arums; and the elongated heart- to lance-shaped leaves of pickerelweeds.

Earlier this spring, the spatterdocks' curled leaves were below the surface, light green in color, looking like lettuce growing on the water's bottom. Now their broad, dark green leaves begin to float on the water's surface and, as the summer progresses, will often stand above the water. The floating leaves, with their long stalks, are connected to large horizontal rhizomes buried in the sediments.

Water Quality

For many years, volunteers have monitored long-term trends in water quality at Jug Bay, including variations in nitrogen cycling. Sampling was conducted at the main river channel (River Pier), three shallow tidal marsh sites (Upper and Lower Glebe Marshes and Otter Point), and a tributary stream (Two Run Branch). The resulting study was particularly interested in looking at the impact of the railroad bed on water circulation and wetland processes.

Over a 15-year period, dissolved inorganic nitrogen compounds (nitrate and ammonium) varied markedly within the wetland system. Mean nitrate concentrations varied in relation to the main river channel, where nitrate levels were highest. Ammonium, which comes from the river and the wetlands, showed less variation. Two Run Branch, which receives nitrate from groundwater and agriculture, has an average level between those in the main river channel and the three marsh sites. Nitrate levels fall as the residence time of this component increases through Upper Glebe Marsh, Lower Glebe Marsh, and Otter Point, respectively. The investigators attributed the differences in concentrations at the marshes to the damming effect of the old railroad bed on the site.

The information from this sampling was of particular value in demonstrating how dramatically nitrate levels changed daily between high tide and low tide across the tidal marsh surface—strong evidence for nitrate uptake by benthic (bottom-dwelling) phytoplankton (diatoms) and emergent plants. This was most evident during the warm months, but such biogeochemical processes also occurred at a significant level in winter, indicating that denitrifiers (microbes in plants that reduce nitrates and produce nitrogen) are active even at low temperatures.

The striking seasonal variation in dissolved inorganic nitrogen underlines the importance of biological processes (uptake, assimilation, and denitrification) in the wetlands that result in lowered levels of the nitrogen.

Source: Swarth 1994.

Spatterdocks and other emergents will provide habitats for many invertebrates—food for herons over the next few months. Wood ducks and soras consume the seeds of these plants, while muskrats and North American beavers feed
on the fleshy roots. These emergents will also offer elegant flowers of yellow,
white, and violet-blue hues. A sight to behold, and one I always celebrate.

13 *April* | TREES MURMUR IN THE WIND AND SPATTERDOCKS ARE RESPLENDENT

Weather conditions are becoming more favorable to early morning activities.
Initial shivering quickly gives way to the maturing sun, and I'm back at Jug
Bay, standing on the Observation Deck overlooking the Patuxent River, which
meanders through an ever-greening marshland and the surrounding woodland and
farmland. Mist gives resonance to the calls of birds and frogs; laughing gulls and
herring gulls chatter in one of the embayments; and an osprey swoops and calls
before landing on a nesting platform to join his sitting mate. Along this stretch of
the river, there are now some 11 platforms occupied by ospreys incubating eggs.

Tree swallows sweep low over the water. One lands nearby, displaying its magnificent iridescent green and blue feathers, alabaster white breast, and very slightly
forked tail. As always, a great blue heron, with its spear-beaked head and priestly
movements, patrols the edge of a reed bed. More migrants are in evidence, with
Forster's terns and Caspian terns passing over the marsh, and a spotted sandpiper
searching in vain for a mud bank now covered by the high tide. A double-crested
cormorant stands like a gargoyle on a moored log.

The greening trees now demand neck-aching searches for migrants, but, along
with my birding colleagues, I hear and then spot a pair of northern parula warblers,
a lone common yellowthroat, palm warblers, and a blue-gray gnatcatcher. On the
forest floor an eastern fence lizard is barely visible on a rotting log, but it choses
to dash for the nearest tree. When pursued, the lizard stays on the opposite side
of the tree, much like a squirrel. Fence lizards mate from April to August and lay
3–16 eggs. The males often do "pushups," flashing their blue throats and bellies to
attract mates or to warn males violating their territory.

In the greening undergrowth, white blossoms on the weaving, curvy branches
of flowering dogwood now stand out, along with creamy flat-topped clusters on
the more upright, multistemmed blackhaw viburnums Within a few weeks, as

these and other understory plants become fully leafed, it will not be possible to see into the woodland.

In more watery habitats, a northern watersnake glides among vegetation at the water's edge. A muskrat uses its tail and webbed feet to propel itself between ever lengthening mud-covered spatterdocks and breaks the surface film now laden with pollen and expended tree flowers. Nearer the shoreline, cattails have reached two to three feet in height, and broadleaf arrowheads and green arrow arums are one foot tall. In a shaded spot I see the oval, coarsely toothed, bluish-green leaves of a dense stand of jewelweeds. I turn over the leaves to see their pale gray or bluish-green or whitened undersides and observe their succulent translucent stems. Dew has beaded on some of the leaves, forming sparkling droplets; hence its name, jewelweed.

Before I head back, I take a few moments to gaze at some turtles basking on a far off log that is stranded in the main channel. Red-bellied or eastern painted? I cannot tell.

I'm told that gizzard shad are moving up the Patuxent River and other Chesapeake Bay tributaries and that a few late-running white perch are still arriving.

15–18 April | Warming in Fits and Starts

Temperatures are oscillating madly: on 15 April it was 70°F; on 16 April, 82°F; on 17 April, 82°F; and on 18 April, 70°F—all against a normal high in the 60s. What will nature make of this?

More butterflies are emerging as the days warm: eastern commas, cabbage whites, and spring azures. I especially enjoy the tiny, metallic blue spring azures as they bounce through the woodland undergrowth and across the marsh.

A northern watersnake has emerged from its burrow and basks in a sunny spot atop a branch in the Beaver Pond, and I am careful to sit at a distance. Due to its highly variable coloration—brown, grayish, tan, or reddish—this snake has been known by some 14 common names. Some of these names reflect the dark cross bands on its neck and dark patches on the remainder of the body, often leading people to mistakenly identify this snake as a species of cottonmouth or copperhead.

Active during the day and night, northern watersnakes will spend the coming months hunting along the water's edge in search of invertebrates, small fish, frogs, and even small birds and mammals. When attacked, it will bite repeatedly, as well as release excrement and a pungent musk.

April to June is their mating time, so I'm not surprised to see other individuals as I scan the pond. Unlike many other snakes, the female carries eggs inside her body and gives birth to live young—as many as 30 at a time—sometime in late summer to early autumn. Hibernation is from October to March, usually in the burrows of crayfish or voles, or on a rock over or near water.

As I get up to leave, the one basking nearest to me dives into the water, causing others around the pond to follow.

15 April | PAWPAW TREES

Why is it that people find so much mystery in pawpaw trees? I happened upon a thicket of individual trees in the now shady forest. Their pale green flowers will soon turn brown, then maroon or purple. Everything about this tree is strange. At 10–12 inches long and 4–6 inches wide, its leaves are among the largest and turn into the mellowest of autumn's tones. The flowers belong to the tropics and ferment with a grape-like odor. The somewhat banana-shaped fruit grows uglier with age and is edible only when fully mature. Its soft, yellow-orange flesh is custardy, which places it in the custard apple family (Annonaceae), with pawpaws and pond apples being the only representatives in North America. The other 600 or so species are confined to the tropics.

Pawpaw flowers attract blowflies and carrion beetles for cross-pollination, and their resulting fruits are eaten by raccoons, gray foxes, opossums, and squirrels. The tree's leaves, twigs, and bark contain natural insecticides known as acetogenins and thus are rarely consumed by rabbits, white-tailed deer, or insects. Zebra swallowtail larvae feed exclusively on the young leaves of pawpaws, thus conferring protection from predation throughout the butterfly's life, as the acetogenins make them unpalatable to birds and other predators.

The earliest documented mention of pawpaws is in the 1541 report of the Spanish de Soto expedition, which found the Creeks, Cheraws, and Catawbas cultivating it east of the Mississippi River. The folksong "Way Down Yonder in the Paw Paw Patch" celebrates the prominence of these trees in the lives of pioneers who settled the valleys of the southern Appalachian Mountains and the greater Ohio River territory. Audubon included the foliage and fruits of a pawpaw in the background for his illustration of a yellow-billed cuckoo, but he was not well disposed toward the plant:

> The fruit, which is represented in the plate, consists of a pulpy and insipid substance, within which are found several large, hard, and glossy seeds. The rind is extremely thin. The wood is light, soft, brittle, and almost useless. The bark, which is smooth, may be torn off from the foot of the tree to the very top, and is frequently used for malting ropes, after it has been steeped in water sufficiently to detach the outer part, when the fibres are obtained, which, when twisted, are found to be nearly as tough and durable as hemp.

The cultivation of pawpaws for fruit production is attracting renewed interest, particularly among organic growers, as a native fruit to be grown without the application of pesticides.

16 April | IN THE FAMILIAR WE FIND SURPRISES

Something about the fencepost caught my attention, and instead of following the path winding through dark trees against the orchestral sweep of an open sky, I picked out an ensemble that gave much delight: the moss, water, and tree rings on a fencepost.

Many times had I passed this way, so what brought the post to my attention this time? Perhaps it was the showery light that made me notice smaller details—more circumspection leavened by a gorgeous spring morning? Maybe the post was of aberrant stock and began to decay earlier than its neighbors. Whatever the explanation, the microclimate, fungi, and microorganisms had gouged a hole sufficient to retain rainwater.

Into this tiny pool had fallen seeds and puce-colored leaves, and around it spread "lawns" of chartreuse-green algae. A cushion of moss looked like a woods seen from above, with the tiny pool reflecting the fleeting sun.

This is a fencepost colorful eastern bluebirds will make joyous in summer, and one from which Carolina wrens will offer an intense volley of song in winter. And then there are woodpeckers that may come to drill.[3]

All this reminded me of the Romans' *genius loci*, the protective spirit of a place or, in more contemporary usage, a place's distinctive atmosphere. In Alexander Pope's *Moral Essays*, his "Epistle IV" (1731), addressed to Lord Burlington, gave full instruction:

Consult the genius of the place in all;

That tells the waters or to rise, or fall;

Or helps th' ambitious hill the heav'ns to scale,

Or scoops in circling theatres the vale;

Calls in the country, catches opening glades,

Joins willing woods, and varies shades from shades,

Now breaks, or now directs, th' intending lines;

Paints as you plant, and, as you work, designs.[4]

22 *April* | SURPRISING MOSSES

Today's presentation on mosses by Dave Davis at Jug Bay's Wetlands Center revealed some very surprising attributes. Everyone present knows that mosses are commonly found in damp, shady locations and perhaps best visualized as carpets on woodland and forest floors. Most of the audience is aware that mosses break down exposed layers beneath the surface soil, releasing nutrients for the more complex plants that are likely to succeed them. Some may recall that individual mosses are usually composed of leaves (phyllids, which are generally only one cell thick) attached to a stem structurally different from the ones for vascular plants (which have tissues to distribute resources throughout the plant). Less well known is the information that mosses are now scientifically classified as the taxonomic division Bryophyta (separate from hornworts and liverworts), and that most mosses have a special mechanism to disperse their spores. In the genus *Sphagnum*, the spores are projected about four to eight inches off the ground by compressed air contained in capsules, thus being accelerated to some 36,000 times the Earth's gravitational force. What came as a great surprise to the audience, however, is that some mosses have proven to be excellent model organisms for analyzing repairs of DNA damage in plant cells. A loss of the ability to repair double-strand breaks and other damage to DNA leads to cell dysfunction, infertility, or death. The determination of the genome sequence of earthmoss has allowed the identification of numerous genes that encode proteins necessary for the removal of damage to DNA.

As mosses dry out, they enclose their cell contents in a sugar gel and synthesize repair proteins. When desert-dwelling mosses were rewetted, these proteins, made earlier, efficiently repaired damaged membranes and chloroplasts (the site of photosynthesis). Complete repair of DNA damage takes about 24 hours, but net photosynthesis can occur in 20 minutes.

After having been told that there are an estimated 15,000–23,000 moss species worldwide, with some 1,400 in the United States and 350–400 in Maryland, we leave the classroom on a rainy and blustery afternoon, confident of finding a substantial number of mosses. There are compact bluish-green pincushion mosses, or leucobryum mosses, on tree bases, pale green tufts of silvergreen bryum mosses occupying disturbed habitats, and dark green, toothed plagiomnium mosses (looking like tiny vascular plants) on streambanks.

We have to await the advance of the seasons to see water moss, also called brook moss or fountain moss—found in flowing freshwater streams and ponds—with its long, slender branches covered with glossy, yellowish-green or dark green leaves. But we do find prairie sphagnums, or blunt-leaved bog mosses, forming carpets in the seepy swamp area of Riggleman Preserve south of the River Farm. Like other mosses of this type, it can soak up water up to 30 times its own dry weight, thanks to its elastic spiral fibers. It grows in wet forests, ditches, and stream margins, but at this time of the year it is just beginning to stir in the warmer waters of spring.

I have always known that birds use mosses for padding and lining their nests, but I had not realized that mosses and lichens possess antimicrobial properties. Such properties are exploited by western bluebirds, pygmy nuthatches, and tree swallows to protect against biodegrading of their feathers.[5]

In Virginia, an examination of the nests of 12 birds from the same area found a total of 65 species of mosses, along with other materials. In the nest of an eastern towhee, the entire inner lining was constructed of many stalks supporting the reproductive capsules (setae) of haircap moss.

Dave is an astrophysicist by profession, but his passion for mosses and help with various Sanctuary activities is truly impressive and is an example to all.

23 April | PONDERING YESTERDAY'S EARTH DAY

Pressures on the natural environment have motivated people to forge new approaches to conservation. Increasingly, gardeners tackle climate change and other global concerns at the local level, as they feel they are contributing to a solution and may claim positive results. Because lawns in the United States cover 42 million acres (three times the acreage planted for irrigated corn), gas-powered machinery and the manufacture of chemical fertilizers add significant amounts of carbon pollution. The application of fertilizers to lawns exacerbates nutrient runoff—some 10 times more than that from farms. Consequently, many gardeners are phasing out these industrial products.[6] Some also plant trees to shade their house and thus reduce energy consumption, as well as grow their own fruits and vegetables.

Such attempts are to be encouraged, but there are other great challenges. Across the country, populations of native insect pollinators have been declining for many years, primarily because much of their once diverse natural habitat has been ceded to vast swaths of monoculture lawns—poor in the pollen and nectar on which they depend. Many of us may prefer lush green lawns, and real estate agents urge their upkeep as a showcase for house sales, but they are the most sterile part of a garden.

Doug Tallamy of the University of Delaware considers backyards to be the last place where wildlife can live. In his seminal book, *Bringing Nature Home: How You Can Sustain Wildlife with Native Plants*, he reminds readers that half of the land area in the United States has been converted to cities and sprawling suburbs, with the result that there are 432 endangered and threatened bird species (200 more than merely a few years ago). Consequently, home gardens have to become sustaining oases, as long as they are seen to have vital ecological roles and are not used simply for decoration and, thereby, reducing populations of insects.

Insects are necessary foods for some 96 percent of terrestrial avian species, especially when the birds are rearing their young. It takes 6,000–9,000 caterpillars to raise a clutch of chickadees. As an example, Tallamy counted caterpillars in his white oak tree and his neighbor's callery pear tree (an ornamental species from China and Vietnam) and found a ratio of 470:1. Though butterfly bushes attract butterflies to their nectar, none can reproduce on this alien Asian species, and azaleas do little to support wildlife. For butterflies, it is more appropriate to plant milkweeds, coneflowers, blackeyed susans, sweetscented joe pye weeds, and blueberries, the latter supporting over 288 species of moths and butterflies.

Tallamy's book is more than a call to install native plants. It underlines the absolute interdependence of wildlife and demonstrates in practical terms how gardens can be transformed into landscapes sustaining the great panoply of our natural world.

As for climate change, extreme heat will destroy fruits and vegetables, and insect

pests are likely to produce three generations instead of one. Weeding in many parts
of Maryland now begins in January, displacing its normal start in March or April.

25 April | VARIOUS BOTANICAL ARROWHEADS AND CONVERGENT EVOLUTION

Arrowhead plants (in the genus *Sagittaria*) tease those attempting to differentiate them. One of the commonest at Jug Bay is the broadleaf arrowhead, or duck potato. Its leaves are similar to those of green arrow arums and pickerelweeds, but with diligence the species may be readily identified. Green arrow arum leaves have two distinct basal lobes opposite their broadly tapering tips. At the end of the leaf petiole (stalk), there are three large, divergent veins—one going to each basal lobe and the third to the leaf tip—from which lateral veins emerge. At the leaf margins, the lateral veins merge with veins running parallel to the edge. While broadleaf arrowheads have similar basal lobes, their venation is very different, with parallel veins curving out above the petiole and meeting in the middle and the extremities of the leaves. The veins of pickerelweed leaves originate from the petiole but arch concentrically into the basal lobes, curving upward to parallel the leaf margins. In contrast to the lone basal leaf of green arrow arums and broadleaf arrowheads, pickerelweeds have petioled leaves (arising from the stems). Spatterdocks are a similar species and are often found with the various arrowhead species, although spatterdocks have broader, more rounded leaves with heart-shaped bases.

Is there an explanation for the arrow-like leaf shapes of arrowhead plants? William Sipple has suggested that the two projecting lobes on arrowhead leaves might serve as a counterbalance for the long projecting leaf tip, thereby preventing the stem from crimping or snapping when water rushes over the stems during tidal flows or floods. Their stability helps the plants absorb much of the energy coming from moving water, thus reducing shoreline erosion and allowing suspended sediments to settle out of the water. Another possibility is that their leaf shape could also confer an advantage when the leaves are exposed to high winds at low tide.

The starchy tubers of broadleaf arrowheads make a nourishing food for both wildlife and humans. Native Americans collected these fleshy roots in the fall for cooking (much like potatoes), made a tea from them to relieve indigestion, and applied poultices of the tubers and leaves to treat wounds and sores.

27 April | GREEN CATHEDRALS AND MIGRANTS
TO MATCH

Steady rain during the past three days restores the normal amount of precipitation for April and propels an explosion of vegetation as I drive through green cathedrals for the biweekly bird count at Jug Bay. It is now 60°F, but a high of 85°F is predicted and, with an overnight wind from the south, we can expect many migrants.

Fog bathes the marsh, and we pick up the raucous calls of laughing gulls from across the barely discernable river channels. The osprey platform in front of the Observation Deck is mirrored in the stilled waters, and a female sits in the nest, with her mate wheeling above, high in the sky. A chick is being tended by one of its parents, a particularly heartening reminder of nature's cycle. A tundra swan in alabaster plumage pokes through the fog, and we wonder why it has not joined others that have headed north. Certainly, the eight double-crested cormorants are in no doubt about a need to migrate as they pass over the reeds, which are now barely visible, as the high tide is covering these emergent plants. The cormorants' relatively short wings, due to their need for effective underwater movement, make their flapping flight the highest energy costs of any bird. The high tide is frustrating the foraging of a spotted sandpiper, whose current ploy is to perch on a branch in the water and grab insects from it. We view it through the flowering dogwood now blooming close to the platform.

Along the woodland trails, showy orchids are in bloom, with their hooded flowers and, yes, their showy, typically bicolored lavender-and-white flowers. Farther along are pink lady's slippers, with two fan-like leaves near the ground from which sprouts a long hairy stalk, bearing a single flower. The flower is yellowish brown to maroon, with a large pink to magenta pouch. In time, the plant will grow to be 6–15 inches tall. False Solomon's seals are sporting their diagnostic long and broad alternating leaves and six white flowers. American holly trees are showing buds, and the stalked white clusters of lyre-leaved rockcresses are resplendent.

In a sunny spot on the side of the Railroad Bed Trail, there is sweet cicely, an introduced species, with fern-like hairy leaves (smelling strongly of licorice when crushed) and attractive, creamy white flowers. Its neighbors are tall blue toadflaxes, with slender, erect stems bearing purple flowers, and a patch of wild rye, showing its single flower cluster. The birders are intent on spotting birds, but Jug Bay's botanist, Cynthia Bravo, joins me in identifying plants and simply enjoying their appearance, including Dutchman's breeches, whose stems grow between the crevices of some shaded, moss-covered rocks.

Back to birds, for, in the meantime, we have heard eastern phoebes, wood thrushes, and ovenbirds and swooned over a magnificent eastern kingbird, an orchard oriole, and a white-eyed vireo. Gray catbirds are passing through in fair numbers, and twittering barn swallows flit upriver. All are evidence of spring migration, but it is the warblers that predominate. In quick succession, in low and high habitat, we see multiple warbler species: common yellowthroat, palm, yellow-rumped (or myrtle), magnolia, yellow, and blue-winged. It is the latter warbler that attracts us, for its face and breast are brilliant yellow, and its wings bear two white bars against a slate blue background. Intermittent basso profundo mating calls of male American bullfrogs compete with birdsong. As Mark Catesby, the English naturalist in whose honor the species (*Lithobates catesbeianus*) is named, observed, "The noise they make has caused their name, for at a few yards distance their bellowing sounds very much like that of a bull a quarter of a mile off."

At the blind overlooking Lower Glebe Marsh, the easeful flights of tree swallows are interrupted by the strong, fast, erratic flying of a Wilson's snipe arcing above us, a bird whose speed has been recorded at more than 60 miles per hour. The word "snipe" originated in India the 1770s, when British soldiers hunted snipe as game.

I return to the Observation Deck to sit and admire the view. The mist has gone and the retreating tide reveals a forest of spatterdocks flapping in the wind. Farther off, the emerging green of wild rice contrasts sharply with the already muddy-colored spatterdocks. Leaves of a nearby northern red oak are already being attacked by gall wasps, for I see many a gall, one to two inches in size, on the tree. Named "oak apples" for their resemblance to this fruit, the galls are outgrowths caused by chemicals injected by the wasps' larvae, which are hatched from single eggs an adult female lays in developing leaf buds. The wasp larvae feed on the gall tissue resulting from their secretions, and the galls get bigger as the larvae grow, pupate, and eventually become adults. When the galls dry out, the adult wasps escape from small holes in the galls. After mating, the female wasp drops to the ground and burrows into the soil. She then injects her eggs into the oak's roots. When the eggs hatch, the larvae eat the roots for over a year. After resting in the pupal stage, the wingless adult females crawl out of the soil and up the trunk, until they find a newly forming leaf.

The forest and its edges are blooming with violets, pawpaws, Canadian service-berries, highbush blueberries, lyre-leaved rockcresses, and flowering dogwoods. American strawberry bushes are about to show their masses of beautiful white flowers.

30 April | HISTORICALLY HOT AND HUMID WEATHER DURING SATURDAY'S CLIMATE MARCH

In a 91°F temperature during the afternoon, climate marchers braved one of the most scorching days recorded in the Washington, DC, area. It is 20 degrees above normal and some 30 degrees warmer than the previous Saturday, ranking among the highest temperatures ever recorded in April and hotter than a typical mid-July afternoon. Equally significant, according to the *Washington Post*, it is 19 days ahead of the average initial 90°F day (17 May), as well as the earliest first 90°F day since 2013 (on 10 April). An overnight low of 70°F broke the warmest levels ever recorded on those dates.

Such exceptionally hot weather is destined to become the hottest April on record—more than seven degrees above normal—and will mark the Washington, DC, area's second record-level warm month in the previous three months, as well as the seventh warmest since 2010.

Oak trees have yet to leaf, their silent branches giving the impression that they are reluctant to let go of winter.

After the putative promises of March, our hearts gladden to the magic of April. The dawn chorus of birdsongs is moving into high gear, emerging flowers have poured forth their fragrance, and a rush of green is storming the land. Frogs and toads squabble over mating partners, and reptiles and bats will have emerged from hibernation. Arriving newcomers join pioneering bird visitors, and the resident avians now face increasing competition. A resurgent April now surrenders to a rampaging May.

Pumpkin ash trees at the edge of Upper Glebe Marsh in late spring.
Rob McEachern

MAY
The Planting Moon

April is promise. May is fulfillment. May is a time when everything is happening, when life rises to a peak. May is the birdsong month.

Edwin Way Teale, *A Walk through the Year*

Nothing is so beautiful as Spring—

. . . .

all in a rush

With richness.

Gerard Manley Hopkins, "Spring"

1 *May* | SCUM: BEYOND SURFACE PATTERNS

Pollen has been particularly abundant this spring, and many a water surface is covered with its grains. They become mixed, and patches of pollen are stretched and folded into each other in swirls that become quite beautiful.

In 1827, while looking through a microscope at particles trapped in cavities inside pollen grains in water, Robert Brown noted that the particles moved through the water, but he was unable to determine the mechanisms that caused this (later known as Brownian motion). Albert Einstein published a paper in 1905 that explained how the observed motion was a result of the pollen being shifted about by individual water molecules. This explanation served as convincing evidence that atoms and molecules, long theorized as the constituents of matter, actually exist.

A paper by T. Mahabale in 1968 gives an account of the spores and pollen in aquatic liverworts, nonfern pteridophytes (simple vascular plants), water ferns, and angiosperms (flowering plants)—a group of aquatic plants forming a biological association of often-unrelated elements. These plants show different degrees of "aquaticism" that determine the dispersal mode of their spores and pollen grains.

In truly aquatic plants, the outer layer of a pollen grain or spore (the exosporium) is simple and thin walled, and the spores within are short lived and germinate quickly, being dependent on water for dispersal. In contrast, in semiaquatic plants, the pollen grains and spores have a thick exosporium, and they rely on insects or wind for dispersal. Some plants, such as water ferns, seem to have become secondarily adapted to an aquatic habitat and have special characteristics and dispersal mechanisms. The floating Carolina mosquitofern is one of the few of ferns that produce two types of spores: large female spores and tiny male spores contained in sporocarps (fruiting bodies). These are able to withstand subzero temperatures, as well as desiccation, enabling populations to overwinter and survive adverse conditions. In addition, vegetative propagation in Carolina mosquitofern and another water fern, floating watermoss, is extremely efficient, and these two species are striking examples of propagation by stem fragmentation, aimed at establishing numerous new individuals. If conditions are favorable, Carolina mosquitoferns can double the area they cover in just 5 to 10 days. These plants usually die back or sink in winter, though they can overwinter in milder climates. In spring, as temperatures rise, the plants begin to grow and move to the surface.

A large number of aquatic angiosperms belong to the group of flowering plants known as monocotyledons. Despite their excellent adaptation to a water habitat, reflected in their vegetative characters, they are quite conservative in the properties of their pollen grains and their pollinating and dispersing mechanisms.

In another study, Karen Jensen of Denmark found puddles and pools in the tropics to be of interest, because the larvae of malaria-transmitting mosquitoes (*Anopheles*) feed on bacteria, fungi, pollen, and other nutrients in the surface layer of the water. When the wind blows, it ripples the surface, mixing the nutrients, so the mosquito larvae have more food. Learning more about how nutrients reach *Anopheles* larvae could help fight the spread of the disease.

3 *May* | PHYTOPLANKTON

High school science students visit Jug Bay at various times of the year to learn about the intriguing world of plankton—the microscopic plants (phytoplankton) and animals (zooplankton) that move and drift through the water column from the surface to the bottom sediments. They come to appreciate how crucial these organisms are to the aquatic food web and to water quality by taking part in PLANS (Plankton, Land use, And Nutrient Studies), a program providing hands-on experience for students investigating nutrient enrichment and phytoplankton dynamics in Chesapeake Bay and its tributaries.

While at the Sanctuary, the students cruise on the Patuxent River, analyzing water quality and using plankton nets to collect samples for examination in the lab. They add nitrogen or phosphorus, or both, to the water samples to detect a jump in production of the phytoplankton and algae—a signal indicating which of these nutrients limits plankton growth. The limiting nutrient varies seasonally: spring runoff brings high nutrient loads, followed by nutrient removal and release by sediments later in the year.

Consequently, the students learn that reductions in phytoplankton will only be possible if inputs of both nitrogen and phosphorus are lessened. PLANS culminates with discussions on management actions to lower the concentration of these nutrients and on what students can do, as good stewards, to reduce nutrient-laden runoff. Excess nutrient loads from agriculture and wastewater treatment plants stimulate algal growth, trigger phytoplankton blooms, and cause oxygen depleted zones in Chesapeake Bay and its tributaries.

Under the guidance of Richard Lacouture of Morgan State University, students set up experiments using one of nature's mechanisms for cleaning up excess algae: oysters. When oysters were plentiful in Chesapeake Bay, they gobbled up plankton, filtering the 18 trillion gallons of bay water over several days. Today's oyster population, which has shrunk to a mere 1 percent of its historical size, takes a full year to filter the volume of water in the bay.

Other studies at the Sanctuary have used biotic indicators as a means to detect pollution in water bodies. But that is another story.

4 May | A Walk along the Marsh Boardwalk and into the Woodland

Summer is in the wings. Only the persistent cold snaps are keeping it from assuming center stage. Along the marsh fringes, two to three-foot-tall cattails are interlaced with green arrow arums and spatterdocks, the latter sporting beautiful canary-colored flowers, borne on lengthening stalks. Near the shore, erect ferns

Catkins and cone-shaped fruits of a smooth alder. *Robert Ferraro*

have begun to cover the muddy bottom, competing with jewelweeds. Sensitive marsh ferns, with their sturdy, broad, almost triangular leaves, hardly resemble ferns. They are already producing beaded, fertile flowering spikes on separate leaves. In a shadier spot stands the more delicate eastern marsh fern, unwinding leaf spirals that reach upward, mingling with those already fully developed and gently dancing in the breeze. Dappled light plays on both water and land. It is a moment to stand and stare.

In mid-February, my attention was drawn to the long catkins of smooth alders, a large shrub (or small tree) with serrated leaves. Now the reddish-green flowers are open, and in some cases oval, dark brown fruits are showing. These persist well into the spring and look like miniature pinecones.

As I climb to the woodland, I am aware of the intoxicating perfumes of mountain laurels, viburnums, white fringetrees, and Japanese honeysuckles filling the air. The blooms of flowering dogwoods are now fading, and blackberry flowers steadily emerge. Underfoot, the delicate white petals and pollen-producing yellow stamens of bloodroots mix with violets, Virginia dayflowers (also called spiderworts), blue toadflaxes, and golden ragworts. Less easy to see are jack-in-the-pulpits and morel mushrooms.

Oak trees produce separate male and female flowers on the same tree, and their productivity has been most pronounced this spring. The male flowers develop first, just as tiny leaves begin to form. The flowers are conspicuous: slim, cylindrical yellow clusters, or catkins, dangling from the tips of branches. As the foliage is sparse, pollen from these flowers drifts relatively unimpeded through the air to pollinate female flowers on nearby trees.

The female flowers are much less obvious, because they are very small and are generally found on the tips of branches higher up in the tree. If pollinated, the female flowers will give rise to acorns.

Once their pollen is shed, the catkins dry up and drop, and the ground under the trees becomes littered with these spent flowers. At home, I find them on the deck and roof, in gutters and on cars, all surfaces covered with a powdery yellow dust—pollen.

CONTROL OF INVASIVE SPECIES

A diverse ecosystem will also be resilient, because it contains many species with overlapping ecological functions that can partially replace one another. The more complex the network is, the more complex its pattern of interconnections, the more resilient it will be.

Fritjof Capra, *The Web of Life*

On this day, a party of schoolchildren has been primed to help remove invasive Japanese stiltgrass by hand, pulling it up before it flowers. After a talk on the significance of their task and a presentation on this nonnative's appearance, the children set off to attack patches of the "offending plant." Their energy is impressive, and their teachers need only encourage some of the less inclined. The accompanying adults tell me most of the students (15-year-olds) enjoy the physical labor as much as being outdoors. The pace and excitement continue, with a luncheon break to refuel.

Japanese stiltgrass was introduced into the United States in Tennessee around

Garlic Mustard

Garlic mustard is a cool-season biennial herb with stalked, triangular to heart-shaped, coarsely toothed leaves that emit a garlic odor when crushed. First-year plants appear as a round cluster of green leaves close to the ground that remain green through the winter and develop into mature flowering plants the following spring. The plants reach two to three-and-a-half feet in height and produce button-like clusters of small white flowers, each with four petals in the shape of a cross.

Beginning in May (in the Mid-Atlantic coastal plain region), seeds are produced in slender pods that become shiny black when mature. By late June, when most garlic mustard plants have died, only erect stalks of dry, pale brown seedpods remain.

Garlic mustards pose a severe threat to native plants and animals in forest communities in much of the Eastern and Midwestern United States. Many native wildflower species that complete their life cycles in the springtime (e.g., Virginia springbeauty, wild ginger, bloodroot, Dutchman's breeches, hepatica, toothwort, and trillium) occur in the same habitat as garlic mustard. Once introduced, garlic mustards outcompete native plants by aggressively monopolizing light, moisture, nutrients, soil, and space. Wildlife species dependent on these early-season plants for their foliage, pollen, nectar, fruits, seeds, and roots are deprived of these essential food sources when garlic mustards replace them. Humans are also shortchanged from seeing a vibrant display of beautiful spring wildflowers.

The pull-and-eat gourmet luncheon organized by the Friends of the Sanctuary often includes garlic mustard.

1919, escaping into the natural environment as a result of its use as a packing material for porcelain. It has subsequently spread to 26 states, occurring on streambanks, river floodplains, and other disturbed areas, growing in full sun to deeply shaded conditions. It is associated with moist, rich soils high in nitrogen.

An invasion by stiltgrass can reduce the growth and flowering of native species, suppress insect communities, slow plant succession, and impair nutrient cycling.

White-tailed deer and other native fauna do not eat this grass, and they facilitate its spread by browsing on native species, thus reducing competition for the exotic (nonnative) plant. It is an annual grass, resembling a small, delicate bamboo, and mature plants grow to two to three feet in height. It spreads by both seeds and vegetative means, with a single plant producing 100–1,000 seeds that remain viable in the soil for at least three years.

Other nonnative invasive plants at Jug Bay are also widespread, with many reproducing quickly. They are being actively managed by staff and volunteers and include the following species: tree of heaven (produces chemicals preventing the establishment of nearby plants); porcelain berry (shades out native plants and consumes habitat); oriental bittersweet (smothers and uproots trees); Japanese honeysuckle (kills shrubs and young trees by girdling—winding around—stems and cutting off water flow); purple loosestrife (forms homogenous—same species—stands that restrict native wetland plants); Japanese stiltgrass, or Nepalese browntop (crowds out native herbaceous vegetation in wetlands and forests); phragmites, or common reed (spreads aggressively and extensively); and Asiatic tearthumb (blocks light and distorts the stems and branches of the plants it covers). In contrast, the aquatic plant hydrilla does not appear to crowd out native species and provides winter food for waterfowl.

5 *May* | Lapse in Summer Feeling

All day it poured, the heavy squalls pelting newly leaved trees and shrubs. Northwesterly gusts rippled the water, and the temperature dropped to the low 60s during the day (a good 10 degrees cooler than the norm) and to the lower 50s at night. By the end of the day, total precipitation brought the level for the year to just short of normal.

I'm told by Greg Kearns that osprey hatching is some two weeks behind the norm, due to inclement conditions. Nonetheless, spawning by mummichogs has been occurring at high tide, especially when the moon is either new or full and when high spring tides coincide with nighttime. They adhere their eggs to plants, such as cattails, where the leaves join the stem, thereby protecting the eggs from being eaten by predators, washing away, or becoming covered in silt. Mummichogs reach sexual maturity during their second year and live for up to three years.

7 *May* | INTERNATIONAL DAWN CHORUS DAY

What sounds to be awakened by!

Henry David Thoreau, *Journals*, 4 June 1852

This day is the worldwide celebration of nature's spectacular choral symphony. Held on the first Sunday in May, it began in the 1980s when the UK broadcaster and environmental champion Chris Baines invited everyone to attend his birthday party at 4:00 am, so that all could enjoy the "dawn chorus." Since then, the event has become an annual international celebration, with over 80 countries now participating. The male chorus happens every day, but springtime is best, when birds nest and advertise for a mate. The songfest is different in each habitat. While listening to it in bed is delightful, it does not compare with standing in deep woodland or on the edge of a marsh.

To maintain the tradition, I rise very early to greet the morn and enjoy one of nature's daily miracles. Like any choir, it's the chorus that predominates: doves, American robins, wrens, northern cardinals, chickadees, and warblers, with occasional solos from ring-necked pheasants, herons, and blue jays. If you are lucky, wood thrushes or hermit thrushes might be nearby, the latter's song being described by Walt Whitman, in "When Lilacs Last in the Dooryard Bloom'd," as "the sweetest, ripest hour of the day."

If you miss the morning serenade, it will strike up again just before dusk and yield better performances from the thrushes.

In general, only the males sing, communicating "keep out of my territory" to rival males and "come hither" to interested females. But why the early start? An obvious answer is that dawn is the best time for the projection of sound. There's usually little wind and less background noise, so singing then can be especially effective. Another explanation suggests that poor light makes feeding difficult—although low temperatures keep insect prey sluggish and within easier reach. Singing at dawn also enables males to announce that they've survived another night and are still defending their territory.

Like humans, birds have regional dialects: members of the same species in different parts of the country have developed striking variations of the same song.

For example, if you listen to song sparrows in various parts of the United States, you would hear dramatically different versions.

The vocalizations of birds in towns and cities have higher pitches than those in the country. This enables their calls to better bounce off the hard surfaces of urban areas. Noise and light pollution have affected the times when birds sing, and studies suggest that climate change has brought new birds, with new melodies, to many parts of North America.[1]

Every day during spring and early summer, a wave of song rises on the Atlantic Coast to salute the advancing day and rolls across the continent, until it eventually breaks and dies away on the shores of the Pacific Ocean. It's one of life's marvels, lifting our spirits as we awaken each morning.

9 May | THE STATE OF CHESAPEAKE BAY

The Chesapeake Bay Report Card released by the University of Maryland Center for Environmental Sciences gives the bay an overall health rating of "C" in 2016. Improved fish populations and water quality contributed to the second-highest grade the ecosystem has received in 30 years. Scientists cite the results as proof that efforts to clean up the estuary are working.

Though the Patuxent River is given a "D+" because of its poor ecosystem health, it is considered to have improved, due to declines in the total amount phosphorus, increases in dissolved oxygen, and lower chlorophyll scores. Clearly, there is much work to be done.

11 May | MIGRATION: COME WHAT MAY

Warblers come north as the leaves unfold. They feed on the forest caterpillars that feed on new green leaves. Their northward flight keeps pace with the unfolding bud and expanding leaf.... Buds burst, new leaves unfurl, larvae hatch, and warblers appear.

Edwin Way Teale, *North with the Spring*

Before the rains come, the bird count group dons waterproof attire in the parking lot as they catch the songs of a scarlet tanager (an American robin with a sore throat), a red-eyed vireo (monotonous American robin–like phrases), an

Wood Warblers

Of the 113 or so species of warblers occurring in the New World, more than 50 migrate to North America to breed. These colorful little songsters, weighing only an ounce or two, usually arrive in early May, as the trees and foliage are leafing out. They migrate at night, stopping whenever dawn or storms prevail. Their breeding grounds offer more food for raising their young and prove to be less stressful than the tropics, where competition is more marked.

The most numerous are the 57 species known as wood warblers: small, active insectivores. Those consuming food on the ground are invariably drab tan or brown, while those feeding higher up are colored bright yellow and black. Yellow warblers, northern parulas, American redstarts, prothonotary warblers, common yellowthroats, ovenbirds, and Louisiana waterthrushes arrive at Jug Bay in early May; other species just pass through on their way to their nesting grounds.

Their arrival as the trees leaf out is timed to coincide with the emergence of many insects, especially in the form of caterpillars. Studies by Robert MacArthur have shown that the part of the tree that is explored (the niche) is quite specific to each species. On coniferous trees, some feed at low- and midlevels (myrtle, or yellow-rumped, warblers), some on the trunk and branches (bay-breasted warblers), some at midlevels near the outside (black-throated green warblers), and others at the uppermost portion of the tree (Cape May and blackburnian warblers).

As the populations of many migrant species are dwindling, it is critically important that the causes are identified and effective measures taken to halt and reverse these declines.

eastern phoebe (a clear "phoe-be"), a blackburnian warbler (a series of high notes, ascending in pitch), and an Acadian flycatcher (a sharp, loud "wee-see"). These are newly arrived migrants, greeted with great warmth and affection by all birders. The male blackburnian is a particular favorite, because of its flame orange head and throat ("firethroat") against a black-and-white body and its dancing behavior high in the tree canopy. Yet its thin, wiry song disappoints.

A quick view of the marsh, with waving spatterdocks and now tall cattails against the darkening and rippling waters, produces numerous tree swallows and the occasional barn swallow. Double-crested cormorants are still heading north, along with two lone Canada geese.

As we quicken our pace through the woodland, we hear wood thrushes and a distant pileated woodpecker. Yellow-rumped warblers, yellow warblers, and blue-gray gnatcatchers are heard and seen, though often hidden by the new leaves.

The rains come, and we make a hasty retreat, but not before seeing a northern waterthrush along the Railroad Bed Trail. Though classified as a warbler, it looks and acts more like a small thrush: streak-breasted and tannin-colored above. It teeters like a spotted sandpiper, picking up food from the ground.

Expert opinion suggests that migration this spring is delayed and "all over the place."[2] Given the effects of climate change, migration specialists wonder if this "strung out" episode will become a pattern, upending a constant phenomenon of many thousands of years.

13 May | INTERNATIONAL BIRD MIGRATORY DAY

Spring is the natural festival of life, and birds are undoubtedly the stars of the show, so it is not surprising that their migration is celebrated today. International Migratory Bird Day (IMBD) is designed to emphasize the importance of conserving migratory birds and their habitats throughout the Western Hemisphere. It highlights and commemorates the journey of nearly 350 species of migratory birds between their nesting habitats in North America and their nonbreeding grounds in Latin America, Mexico, and the Caribbean. IMBD was created by the Smithsonian Migratory Bird Center, and each year it explores a different aspect of migratory birds and their conservation.

Since 2006, people have organized public events—such as bird festivals, education programs, exhibitions, and bird-watching excursions—to celebrate IMBD. This year its theme is "Stopover Sites: Helping Birds Along the Way," and the Friends of Jug Bay have organized a series of presentations on migratory pathways (Matt Jeffrey, National Audubon Society), the value of bird banding and monitoring (Mike Quinlan, volunteer, Jug Bay Wetlands Sanctuary), and initiating studies at the local level, such as those on soras (Greg Kearns, Patuxent River Park).

The presenters stress that when birds migrate between their nesting and wintering sites, they rely on resource-rich and strategically located stopover areas to acquire the fat stores needed to provide sufficient energy for them to fly thousands of kilometers across continents and oceans. Some stopover sites are well known—such as along the Gulf Coast of Louisiana and Texas and the Atlantic Coast—where birds pause after traveling along the shoreline. Other sites are inland, such as Midwestern grasslands, Appalachian forests, Chesapeake Bay wetlands, and even urban parks and backyards. Some species, such as wood thrushes, which have recently arrived at Jug Bay, fly over 2,500 miles from Central and South America. They will nest and raise their young in the Jug Bay region and throughout eastern North America.

As Bruce Beehler notes in his book *North on the Wing*, given that the migrants' instinct-driven seasonal movement from one continent to another is a continuing annual cycle, they adopt a lifetime solution to the ecological challenges posed by living in a seasonably variable world. So it's not surprising that each IMBD presenter emphasizes the need to better understand migration and address threats on breeding and nonbreeding grounds, as well as on stopover sites. They discuss new technologies to track migrants, conservation efforts along the migration flyways, and the importance of protecting habitat at the local level.

A study published two days after the meeting by Mayor and colleagues demonstrates that climate change is altering the seasonal clock on which North American migratory songbirds depend to mate and raise their offspring. A growing shift in the onset of spring has left 9 of the 48 species of songbirds that were studied unable to reach their northern breeding grounds in time to produce the next generation of fledglings.

"The birds are trying to keep up with the speed of climate change but they can't,... it's just too fast," said Morgan Tingley of the University of Connecticut, one of the study's authors. "They can't adapt quickly enough."

To produce healthy young, migrants must arrive at their breeding grounds when the early-season boom in insects coincides with springtime plant growth. But as climate change shifts the timing of leafing out (a temperature-driven process known as green-up), the migrants are more likely to reach their breeding grounds either when temperatures are still frigid and food is scarce or after insect numbers have begun to dwindle.

The study team found that green-up is beginning earlier in eastern North America and—surprisingly—later in the West. Birds that breed primarily in eastern temperate forests tended to lag behind, while species that breed in western forests reached their breeding grounds too early.

Although the team observed migrant North American birds failing to keep pace with the rapidly shifting interactions between vegetation and climate, the specific demographic and ecosystem consequences of these trends remain unknown. Population-based developments are predicted by trends in climatic suitability, and if birds are unable to keep pace with their changing environment, reduced fitness for individuals, decreased population sizes, and—in the extreme—extinctions could result. Not a pretty picture.

14 *May* | SUNSHINE CHASES THE RAIN AWAY

 After three consecutive days of almost constant rain, this day sees clouds sailing across the sky and warm midday sunshine. Some areas experienced more than two inches of precipitation, which broke the drought, bringing the total for this year to just shy of the norm. Breezes and windy gusts (some up to 40 mph) from the northwest kept migrants at bay. This may well change soon, with southwesterly winds and temperatures in the lower 90s expected by midweek.

I make a brief visit to the sandy meadow and observe that it is in full bloom, showcasing many species: tropical-looking longbranch frostweed (in icy conditions, the stems exude water that freezes into fascinating shapes, hence its common name), beautiful blue toadflax, tiny thymeleaf sandwort, and, springing from a rosette (a round cluster) of leaves, mouseear cress, with delicate yellow-white flowers. This last was the first plant to have its genome sequenced, and it is a popular tool for investigating the molecular biology of plant traits.

16 *May* | BIRDS EVERYWHERE, AND CICADAS?

all the merry little birds are
flying in the floating in the
very spirits singing in
are winging in the blossoming

e e cummings, "sweet spring"

The southwesterly winds have helped bring in more warblers and other migrants. Their calls echo in the landscape, and feathered bodies flit through arcades of leaves and branches. The migrants, adorned by bright breeding plumages, now return to forests, grasslands, and wetlands, having flown vast distances. What incredible creatures they are. On one spring day in Wisconsin, I recall having seen some 20 species of warblers in just one large tree. Alas, such a spectacle is no longer possible, given the degradation of their breeding and wintering grounds and the decline in stopover sites.

Much as I like warblers, there is something wholly uplifting about seeing the first barn swallows swooping over the marsh. Twice a year, they undertake the perilous journey to and from South America, so their lives become endless summers.

Every 17 years, we expect cicadas to engulf trees in our yards and woodlands. Now, however, from northern Virginia to north of Baltimore, these red-eyed, barrel-shaped insects have been observed crawling out from beneath trees some four years early. The numbers are not overpowering, but some suspect that this is another demonstration of climate change. Consequently, the many reports of these early emergents (stragglers) concern scientists, who are asking people to continue their observations for the next month and recall how they compare with past cicada eruptions. This will help determine if the longer growing seasons linked to climate change may have shortened the 17-year cycle. Ongoing studies by Chris Simon, an ecologist and evolutionary biologist at the University of Connecticut, have suggested that many more individuals could come out four years early and eventually become so numerous that they would self-reproduce.

Early cicada emergence may be a way to ease high-density populations or expand their population and territory. It may also act as a hedge against natural events. In answer to why cicadas have 13- or 17-year life cycles, it is suggested that they thereby avoid gaining a specialized predator—and no animal has, so far as we know, evolved to specifically prey on these periodical cicadas.

18 May | A Nestling: Northern Cardinal

A disturbance in the water catches my eye and I move closer, expecting the bobbing head of a turtle or snake. But no, it's a young bird struggling to reach dry land. I edge a twig under its body and, with wings much a'flapping, it quickly gains a foothold. It has enormous goggling, toad-like eyes and the beginnings of feathers bursting from sheaths, forming purplish orbs at its sides. The nakedness of its belly, shanks, and neck strike a contrast, but it is the enormous red gape of its mouth that truly impresses. The fledgling must be some eight or nine days old and pulses with life as it clings to its safe haven, crying out for its parents.

I move to a nearby shrub and begin a search for its nest or for signs of a parent. I find no nest, but I do see a female northern cardinal approaching from the undergrowth, clearly concerned by the cry. The half-feathered fledgling calls again, but the parent moves no closer.

I gently lay the twig bearing the terrified bird in a secluded patch of leaves, coming as close as I dare to the parent, and tiptoe away. Happily, parent and young are reunited, with food being taken in with heartening gusto. I stay a while and witness further feeding, but I have to wonder, with nightfall coming, if this baby bird's entry into the world will be short lived.

19 May | Unseasonably Hot Weather

Winds blow from the southwest, and the afternoon temperature reaches 93°F.

25 May | Much Splashing in the Spatterdocks

Furious splashing amid the spatterdocks draws the attention of visitors. The stems sway mightily, the waters churn, and a rushing sound crosses the marsh. Someone suggests otters; others, muskrats. In fact, it is the spawning of common carp, timed for when the water temperature reaches 63°F and high-tide waters rise (as on this day), and when there has been heavy rain (as has happened over three of the previous days).

Several males may spawn with a single female, who can release between 100,000 and 500,000 sticky eggs over a four- to six-week spawning period. The eggs are dispersed by the splashing and adhere to submerged vegetation. They do not receive parental attention and hatch in four or five days. The young fish usually remain in shallow waters, feeding on small crustaceans and grow some three to four inches in their first year of life.

Common carp were brought to the United States in 1831 and introduced to the Chesapeake Bay region in 1877. By the late nineteenth century, the government distributed them widely throughout the country as a food fish, following pressures from increasing waves of immigrants, who could not accept that the vast continent had no carp. A steep decline of native fish stocks made their plea more compelling.

By the early twentieth century, the introduction of carp was so successful that many came to regard the fish as a nuisance, its rapid spread appearing to threaten both water quality and native freshwater species. A more likely explanation is that dredged and straightened channels, drained wetlands, eroded riverbanks, and waters polluted with domestic and industrial wastes had an adverse impact on native species.

Nonetheless, with their large numbers, carp can generate problems. As bottom feeders, they uproot vegetation. After sucking in mud and other matter, they spit out the residual, thereby increasing the water's cloudiness. This impairs the visibility predator fish need to find their prey and reduces the amount of sunlight reaching submerged plants.

Controlling carp has not proven to be easy. Conceding the carp's permanence, removal programs in the 1950s began to concentrate instead on control of both their populations and their migration into game fish waters. There is some hope that an expanding market for carp and carp products could become the long-term check on their population that states' removal programs have not been able to implement, due to limited resources. In parallel, cleaning up the dirty water in which they live could improve the taste of their presently tainted flesh. Immigrants could once again lead the way in this effort.

26 May | Measuring the Health of Bird Populations

I join a team of bird banders who gather every year at this time to participate in the Monitoring Avian Productivity and Survivorship Program, a continent-wide research effort started in 1989 by the Institute for Bird Populations (IBP) at Point Reyes in California. Its goal is to assist in the conservation of birds and their habitats through demographic monitoring, gathering data during the breeding season on the adult population size, survival rates, and breeding productivity of over 150 targeted migrant and resident songbird species. Data collected by the Jug Bay station are analyzed by the IBP to determine how the survival rates relate to such environmental factors as habitat structure, forest composition, and climate change. The Monitoring Avian Productivity and Survivorship Program (MAPS) differs greatly from bird counts, as it involves capturing, measuring, and banding birds, with recapture data helping determine songbird health.

A volunteer extricating a captured bird from a mist net. *Chris Swarth*

Sanctuary staff and volunteers initiated the Jug Bay program in 1990, making this bird banding station one of the oldest continuously operating in the United States. Every summer during the breeding season, lasting from late May through early August, 14 mist nets in the study area are used to capture songbirds to determine each bird's age and sex. Standard measurements are made of the wing cord (taken with the wing bent at a 90-degree angle) and weight for each bird, and a US Fish and Wildlife Service band is placed on the leg of each bird to identify it, should it be recaptured. By 2015, over 3,103 songbirds and woodpeckers had been banded, involving 66 species (54 were neotropical migrants and 12 were resident species).

The first bird dangling in the mist net is a female northern cardinal, with feathers and feet askew. The leading bander, Danny Bystrak, carefully extricates the bird and notes the presence of a brood patch (a featherless patch, reddened by a network of tiny blood vessels), indicating that she is incubating eggs. He measures her wing length, weight, and general condition. An appropriately sized, numbered metal band is clamped onto the bird's leg, the band number recorded, and the bird carefully released.

The MAPS study at Jug Bay has provided insights into such questions as what factors cause reductions in avian populations; where are problems most acute on breeding or nonbreeding grounds; what drives differences in trends between particular regions or habitats; what defines the relationship between population change and weather, climate, or habitat loss; and what measures can be taken to reverse declines?

In 2013, the oldest recapture was a red-eyed vireo, banded at Jug Bay in 2003 and subsequently recaptured in 2006, 2007, 2009, and 2011. The bird was a minimum of 10 years, 11 months old, setting a new North American longevity record for this species.

In July 1999, banders found an old friend in the mist net—a tufted titmouse, originally banded as an adult on May 1993, making it at least seven years old. While it was adept at evading predators, it had been caught at Jug Bay eight times. Nevertheless, with all of its captures during the breeding season, the bird has never developed a cloacal protuberance (a swelling of the tubes producing sperm, indicating a breeding male) or a brood pouch (indicating a breeding female). It is possible that this bird never nested, and the lack of stress associated with breeding may have accounted for its longevity. The bander, Danny Bystrak, speculated that it might have avoided "the parent trap and led an extended healthy life."

27 May | TULIPTREES

Every spring, I single out tuliptrees. They display a stately beauty and are known for their tulip-like flowers that stand erect on every bough, amid broad, thick, rich green leaves. The glossy blossoms set the whole tree alight with their outer deep, recurved, green sepals and inner petals of yellow green, flared with orange at the base.

I can do no better than to quote Donald Culross Peattie: "But despite the splendor of its dimensions, there is nothing overwhelming about the tulip tree, but rather something joyous in its springing straightness, in the candle-like blaze of its sunlit flowers, in the fresh green of its leaves, which, being more or less pendulous on long slender stalks, are forever turning and rustling in the slightest breeze; this gives the tree an air of liveliness lightening its grandeur."

The leaves have their own beauty. The glossy green deepens as the summer progresses, turning a rich gold by autumn. Then, in winter, the leafless tree stands alone, in a magisterial pose.

28 *May* | AN OBVIOUS CONNECTION: JUG BAY AREA BIRD BLITZ

An Important Bird Area (IBA) is a locale that is identified, using an internationally agreed upon set of criteria, as being globally important for the conservation of bird populations. Jug Bay is classified as an IBA, and a Bird Blitz gets underway, in order to determine whether the surrounding woodlands should be added to the IBA site currently covering the wetland and its waterbirds. This effort is designed to collect data needed to evaluate potential sites, and more than 100 volunteer birders have helped conduct Bird Blitz surveys at candidate sites across Maryland.

Today is a training session to induct volunteers into survey methods, which bird species to record, and data collection. One criterion for becoming a site is predicated on the presence of significant breeding or nonbreeding numbers of species at risk in Maryland. Others require that the locale be an important gathering site for bird species dependent on a particular habitat type, or that native species of birds regularly concentrate there in significant numbers when breeding, either in winter or during migration.

Jug Bay qualified as an IBA in 2004 on the basis of meeting all three criteria for secretive marsh birds. The issue is now whether forest interior dwelling species in the uplands of the Jug Bay area meet any of these criteria. With the recorded breeding of 19 such species, extending the current IBA to embrace woodland should be achievable. Such a gain would have the benefit of better integrating these very different habitats and imposing a management framework for more-effective conservation.

IBAs constitute building blocks to help create a network of intact, resilient, connected natural areas and working lands, thus sustaining healthy and diverse populations of wildlife. They also provide clean water, clean air, flood protection, recreational areas, and an improved quality of life for the people of Maryland—and beyond.

As I walk from the marsh into the woodland, I'm aware of a transition zone (ecotone) blending two very different habitats (marsh plus wetland, and the beginning of upland forest) and possessing the characteristics of each. Plants on both sides extend out as far as they can. Beyond this zone, competitors in the adjacent community take over. This is known as an edge effect, and it is due to a locally broader range of suitable environmental conditions, or ecological niches. Ecotones are particularly significant for mobile animals, and this particular ecotone includes a mixture of adapted species, such as eastern towhees, common yellowthroats, wood thrushes, yellow-throated vireos, prothonotary warblers, and scarlet tanagers.[3]

I then decide to walk up the middle of a small ravine that extends from the marsh into the upland forest. It is a riparian forest ecosystem, whose habitats are dependent on the existence of perennial, intermittent, or ephemeral surface or subsurface water drainage. Biologists, naturalists, and others have long recognized the great value of riparian ecosystems to fish and other wildlife, including populations of various aquatic insects. They are in close proximity to diverse structural features—live and dead vegetation, varying topography, biogeochemical processes, and geological substrates—and thereby provide extensive edge areas and mixed wildlife habitats and corridors, offering pathways for migration and other types of movements between habitats. The abundance and diversity of various songbird and small mammal species are centered on activities related to moisture. For example, moist soils are required by some bird species for feeding (e.g., American woodcocks) or preferred nesting habitats (e.g., prothonotary warblers).

How, I wonder, will the IBA survey capture these ecotones and justify an IBA extension?

31 May | Dragonflies: Creatures of the Sunlight

I walk next to the marsh and note that adult dragonflies have begun to emerge, only to suffer a short life. After a year spent underwater as nymphs, as adults they have only a few weeks to mate and lay eggs. Until research by Don Williamson (a biologist at the University of Liverpool) was published, nymphs and airborne adults were thought to be progressive stages in the metamorphosis of an insect that has a common ancestor and is one animal. His controversial work proposes that the two phases are two different animals.

Williamson contends that under his larval transfer hypothesis, one or more ancestors of damselflies acquired larvae by hybridizing with three-pronged bris-

A dragonfly nymph. *Kerry Wixted*

tletails—small wingless insects moving like fish, with small compound eyes, long antennae, 10-segmented abdomens, and three tail-like structures. Their mouthparts are partly retractable, with simple chewing mandibles, or jaws, and long antenna-like appendages extending from the upper jaw. The best-known bristletails are silverfish, scurrying away from light in a kitchen to dark damp places.

Williamson suggests that the hybrids evolved into a nymph with a bristletail, but also with larger eyes, an expandable jaw to seize water fleas and mosquito larvae, and bristles that have become gills. The nymph molts its hard outer shell many times before the final molt, then climbs out of the water to split open and emerge as an adult damselfly. He notes that there is less difference between insects that hybridized (e.g., bristletails) than between other insects (such as beetles and butterflies) and their grub or caterpillar larvae. Thus, passage into a pupal stage and then a start-again metamorphosis no longer were necessary.

According to Williamson, the organs and tissues of butterfly and moth caterpillars break down in the pupal stage and are replaced by a "soup"—a mixture of liquid, stem cells, and small "imaginal discs," formed during the last caterpillar stage, that develop into the tough outer cuticle, legs, wings, and nerves of the adult growing from them. Other adult organs, including the heart, gut, and digestive gland, grow from stem cells in the pupal soup, and no part of the larva becomes a part of the adult. The developing insect thus dismantles the larva and then starts again to produce an adult. Williamson questions whether "this complex procedure could have evolved by a series of small changes, each subject to natural selection. In contrast, dragonflies and damselflies manage the transition from larva to adult without it."

Williamson's hypothesis has drawn heated debate, with those rebutting it calling for laboratory hybridization studies to provide evidence of evolution through hybridization.

Whatever the final outcome of these studies, I watch an adult common white-tail wriggle from its own skin, rest on the leaf of a cattail, and pump a bodily fluid, known as haemolymph, into its limp wings, which slowly expand by its sides. Over the next few hours or days, depending on the temperature, the dragonfly's exoskeleton hardens, and it will fly off in search of flies, mosquitoes, and other small insects. Its colors will become more vivid in a matter of days.

And so we pass to June—May having rapidly applied splashy colors onto an ever-expanding canvas—a month that appears no longer hesitant, but glossy and luxuriant. Scents are buoyed on warming nights under a brilliant moon. Mammals see their young bounding across the landscape, while those that had hibernated are pressed to catch up. An outburst of invertebrates will stoke the hungry mouths of chicks. Spring has reached high tide and inexorably cedes to summer in one great symphony.

An early summer view of Upper Glebe Marsh, with swamp chestnut oaks, from the Observation Deck. *Jug Bay Wetlands Sanctuary*

JUNE
The Rose Moon

June comes with its own tranquility, predictable as sunrise, reassuring as the coolness of dusk.... There is certainty, an undiminished truth, in sunlight and rain and the fertility of the seed. The fundamentals persist.

Hal Borland, *Sundial of the Seasons*

1 June | WHITE-TAILED DEER: BAMBI REVISITED

Female white-tailed deer are now giving birth, and the fawns, in their reddish-brown coats with a scattering of white spots (helping camouflage them), are a stunning sight. Their teddy-bear ears bat back and forth to catch the slightest sound. The mother nurses her fawns, a practice she will continue until they are 10 weeks old, when they will add vegetation to their diet. And therein lies the rub!

Over 70 years ago, white-tailed deer numbers were held in check by subsistence hunting and, before that, by those at the top of the food chain that have no natural predators, now gone. But with the advent of hunting regulations, their populations began to increase, so that by 2004, these deer were considered overabundant in 73 percent of their range in eastern North America, due to their devastating browsing on the forest understory and disruption of the cycling of nutrients and energy in food webs. Songbirds dwelling in forest interiors need appropriate nesting places and rely on insects, worms, and other invertebrates for sustenance, so they are especially sensitive to changes in this habitat caused by overbrowsing white-tails.

In Virginia, biologists Thomas Rooney and Donald Waller have found that study sites with high white-tailed deer populations were virtually devoid of hooded warblers, prairie warblers, and white-eyed vireos—three species that rely on a forest's undergrowth. Many populations of these avian species are already decreasing, due to habitat loss. Prairie warblers, for example, have been designated as a species of conservation concern, following range-wide population declines.

In Connecticut forests, the populations of many songbirds, including worm-eating warblers, black-and-white warblers, hooded warblers, ovenbirds, ruffed grouses, eastern towhees, and brown thrashers, have declined significantly. White-tailed deer are also eating oak, hickory, and hemlock saplings—trees unable to regenerate, given the repeated damage they have suffered. Natural succession and biological diversity are in decline, and invasive plants are spreading. Consequently, bird species dependent on these trees, such as eastern wood-pewees and wood thrushes, are disappearing. Wild flowers, such as Virginia springbeauties, showy orchids, bloodroots, and other species, have become rare.

Because of these losses of understory and bird species, many conservation organizations and governmental agencies are taking steps to selectively reduce, or cull, the deer herds. By monitoring the understory and keeping tabs on bird populations, biologists will have better measurements to effectively regulate deer numbers.

At Jug Bay, a deer management program was implemented in 2009, with significant numbers of white-tailed deer being removed from the Sanctuary and nearby Glendening Nature Preserve each year. Data collected on the forest understory since that time records a doubling of the species diversity of plants for both deer-excluded and deer-accessible plots, with a significantly higher number of seedlings found in deer-accessible plots than those excluding deer.

Though these findings are encouraging, of the 21 tree species studied over a seven-year period, only six have succeeded in reaching the sapling stage without the protection of exclosures. Should this trend continue, the forest may be subject to a shift in species dominance and a loss of species diversity. Consequently, there is a need for such deer management to intensify.

The white-tailed deer turn and bound away as if they have to negotiate gaps in the air. They present a beautiful sight, melting into the dappled light.

2 *June* | THE RAILROAD BED TRAIL

With fragmented thickets of honeysuckle on either side of the railroad bed, I can now appreciate why this section became known as the "Honeysuckle Route." Trains carried vacationers from Seat Pleasant, on the Washington, DC / Maryland boundary, to the resort town of Chesapeake Beach on Chesapeake Bay. The route was finally completed in 1900, but its trains only ran for 35 years.

One of the obstacles faced in constructing the Chesapeake Beach Railway was crossing the Patuxent River. A wooden trestle was preferred, but steamers coursed up and down the river twice a week to carry tobacco, lumber, and other cargo from wharfs at Pig Point.

A solution was to build a movable plate girder swing span, supported by a cylindrical concrete base in the middle of the channel, leaving the water deep enough for the passage of river-borne traffic. On the Anne Arundel County side, a quarter-mile-long causeway brought the route's single track to the swing span, and a 300-yard trestle did the same on the Prince George's County side.

In 1935, the swing span, its rotating mechanism, and the 300-yard trestle were removed, leaving the cylindrical base still visible to this day. Recent storms have caused extensive damage, and one wonders when this remnant of the Chesapeake Beach Railway will slip beneath the waters and be lost to history.

3 *June* | INSECT ATTACKS ON LEAVES AND DEFENSIVE MECHANISMS

At this time of the year we become conscious of insects invading our homes, as well as their buzzing and murmuring in trees and other vegetation. Many of the countless insects living on, in, and around trees are harmless; others can inflict great damage, sometimes leading to the trees' death. We see holes in leaves created by chewing insects; tunnels in roots, stems or trunks, twigs, and leaves by boring insects; and damage to twigs and leaves by sucking insects.

You might conclude that plants are entirely passive, but this is far from the case, for they are good at sensing an herbivore. For example, studies on mouseear cress, or thale cress (an invasive from Eurasia), have shown that it can pick up vibrations from feeding insects and boost the level of its bitter-tasting defensive chemicals to ward them off. Botanists Heidi Appel and Rex Cocroft have found that the plant not only detects feeding vibrations, it responds selectively to them. They then posed the question, how does the plant raise the alarm? Edward Farmer has discovered that when a leaf is cut or wounded, the plant uses electricity to buzz a warning to other parts of the plant. He goes on to identify two particular genes needed to initiate the buzz, finding that they are closely related to genes in synapses of the human nervous system. A guess is that two nutrient transport systems within the plant are involved in responding to herbivory: phloem and xylem, working together somehow to convey electrical signals.

Defensive mechanisms of plants against herbivores are proving to be wide ranging and highly dynamic. They include structural barriers, toxic chemicals, and the attraction of natural enemies of the target pests. Structural traits—such as leaf surface wax, thorns or surface hairs, and cell wall thickness—form the first physical barrier to feeding by herbivores. Biochemical mechanisms involve defensive compounds either stored as inactive forms or produced in response to plant damage. These adversely affect the feeding, growth, and survival of herbivores. In parallel, some plants also release volatile organic compounds that attract the natural enemies of the detrimental herbivores. These strategies either act independently or in conjunction with each other. The production of these readily vaporized substances and the secretion of extra floral nectar facilitate the interactions of plants with the natural enemies of insect pests—such as parasitoids, which live within a host and eventually kill it—or predators, and actively reduce the numbers of feeding herbivores.

4 June | "Sumer Is Icumen In"

Clouds, rain, and grayness have given way to sunlight dancing in the trees and the undergrowth. Beguiling morning sunrises and dazzling sunsets encourage us to slow down, take prolonged walks, and enjoy the rhythm of the season. Phragmites and common rushes wave, while thin-leaved sedges lurk on the marsh's edge. Trees are fully leafed, and a languorous landscape moves toward summer. Languorous as a whole, perhaps, but in detail, nature, as Alfred, Lord Tennyson, observed in his poem "In Memoriam," is "red in tooth and claw." As I pass Beaver Pond along the Two Run Trail, two turtles are lurching mightily (possibly snapping turtles, for they have lightning reflexes), one having seized the neck of its rival. The tumult persists for some minutes, until they separate and dive underwater.

American bullfrogs show their large protruding eyes amid common duckweeds, and the plucked-banjo-string calls of green frogs drift across the floating and emerging vegetation. A few northern watersnakes and red-bellied turtles bask on logs lodged at the pond's edge. For a moment, the distant call of a woodpecker competes with songbirds in the nearby treetops.

World Environment Day

Every year, World Environment Day is celebrated on 5 June. First held in 1973, it has become the United Nations' principal vehicle for encouraging an understanding of and actions for the protection of our natural environment. It helps raise awareness of emerging environmental issues, ranging from marine pollution, human overpopulation, and global warming to sustainable consumption and wildlife crime. World Environment Day has grown to become a global platform for public outreach and adopts annual themes that major corporations, nongovernmental organizations, communities, governments, and celebrities worldwide embrace to advocate for environmental causes.

The theme for 2017 was "connecting people to nature—in the city and on the land, from the poles to the equator." Canada was the host nation, and World Environment Day was a component in that country's 150th birthday celebrations. As part of the festivities, Canada offered free passes to its national parks throughout 2017.

Billions of rural people around the world spend every working day "connected to nature" and increasingly appreciate their dependence on potable water supplies and the ways in which nature provides their livelihoods. They are among the first to suffer when ecosystems are threatened, whether by pollution, climate change, or overexploitation.

Nature's gifts are often hard to assess in monetary terms. Like clean air, they are often taken for granted, at least until they become scarce. In recognition, economists are developing ways to measure the multitrillion-dollar worth of many ecosystem services, from insects pollinating fruit trees to the leisure time, health, and spiritual benefits of nature.

World Environment Day is an ideal occasion to go out and enjoy national parks and other wilderness areas. Park authorities in some countries may follow Canada's example and waive or reduce park entry fees on 5 June or for a longer period.

By now, most of the migrant birds have passed through. The breeding season of resident birds is well underway, with the calls of nestlings permeating the understory. More young birds are around, and a vortex of gnats in the rising heat invites swallows to an easy meal.

It is a beauteous time, a time when we open doors and commune with nature. Wordsworth's "living air" unleashes the frontal cortex of our brain and frees us from heavy, persistent thoughts. Nature becomes a place with no burdens and no pressures, where one is never fully alone.

11–13 June | Heat

Bright, sunny, decidedly hot weather arrived on 11 June, along with summertime mugginess and, at 90°F, temperatures some five degrees above the norm. This will continue for the next two days.

12 June | Amber Marsh Snails

There they are in all their glory: delicate small snails with a glossy, amber-colored shell, embellished with an S-shaped stripe revealing its digestive tract. I'm in the high marsh, amid cattails, green arrow arums, jewelweeds, and halberdleaf tearthumbs, for I know these snails are not an aquatic species: they lack gills enabling other creatures to stay underwater.

I observe the snails' eggs, which adhere to plants, algal mats, or other sites reached by water only at high spring tides. Embryos will develop when exposed to moist air and hatch after 7 to 10 days, when such tides reach them. Hatching is initiated by a lack of oxygen, occurring when a boundary layer of relatively still water surrounds the metabolically active egg at high tide. The young spend their first few days in shallow pools at the base of marsh plants, as protection from predatory fish. Both juvenile and adult snails feed with radulas (a rasp-like structure composed of tiny teeth), scraping pollen and detritus off the plant stems on which they live.

Unpublished studies at Jug Bay undertaken by Christopher Swarth, Michelle Reynolds, and Kathy Szlavecz show that amber marsh snails' densities peak in mid-July. Their findings suggest that the absence of snails on wild rice and phragmites may be due to silica in the plant cells, making digestion difficult. The snails are also not present on spatterdocks, due to these plants' low position in the marsh,

Noises Under Water

As a child, and later as an aquatic ecologist, I was drawn to whirligig beetles and water striders effortlessly plying water surfaces, and to water boatmen and giant diving beetles swimming below. But only now have I learned that these and other water denizens produce sounds that are the loudest in the animal kingdom, relative to their body size. Scientists have recorded water boatmen "singing" at up to 99 decibels, the equivalent of listening to a loud orchestra.

The 5–15 millimeter-long lesser water boatman makes this sound by rubbing its penis, a stridulatory peg, over a plectrum, creating a pulse of sonic vibrations. The more a peg is scraped over the plectrum, the more variable the pitch, intensity, and rhythm of the sound. Researchers say the song is a courtship display performed to attract a mate, and sexual selection could be the reason why the insects' songs reach such a high amplitude. "We assume that this could be the result of a runaway selection," biologist Jérôme Sueur from the Museum of Natural History, Paris, told the BBC. "Males try to compete to have access to females and then try to produce a song as loud as possible, potentially scrambling the song of competitors."

What makes water boatmen extraordinary is that the area they use to create sound only measures about 50 micrometers across, roughly the width of a human hair.

Source: Sueur, Mackie, and Windmill 2011.

where hatchlings are threatened by drowning during high tides, despite their being able to move at 0.4 inches per hour.

To add to their unusual nature, amber marsh snails manage to endure winter by burrowing into the mud, where they survive underwater, in spite of a lack of atmospheric oxygen.

14 June | TREES: AN UNDERGROUND SOCIAL NETWORK?

Off and on, I have been reading an extraordinary book, *The Hidden Life of Trees,* by Peter Wohlleben, a German forester. The author presents trees as a networked, intentionally collaborative, talkative community. They "experience pain and have memories ... [and] parents live together with their children." They can behave well or badly. Some bully. Others grasp, are wasteful with energy, or are frantically impatient to grow. Trees are aware of decreasing temperatures in the autumn and can "compare day lengths ... and count warm days in the spring." Should a neighboring tree fall down, others close by "suffer the temptation to do something stupid" by growing a branch in the new space. It's "an individual choice and, therefore, a question of character."

Wohlleben goes on to claim that trees "pass on their knowledge to the next generation" by teaching and learning. In other words, they are social agents, with a high level of control over how their bodies grow and function. The latter action is mediated in the sensitive, searching filaments of thousands of root tips or by chemical messages fanning through the forest floor via symbiotic fungal mycelium (the vegetative part of a fungus, living in a close physical association with the tree's roots).

Wohlleben cites studies demonstrating sensory alertness in leaves and giving examples of caterpillar attacks that stimulate electrical signals, prompting the release of defensive compounds. Trees broadcast to other trees, helping them preempt similar attacks.

To support the author's claims, the book ends with a note by Suzanne Simard, a professor of forest ecology at the University of British Columbia. Using sensitive, specialized measuring devices, she found "carbon being transmitted back and forth between the trees, like neurotransmitters firing in our own neural networks. The trees were communicating through the web!"

16 June | THOSE WHO WISH TO SING WILL ALWAYS FIND A SONG

Birdsong is now at its best, inviting reflections on its range and whether it may be judged as music. Songs are most highly developed in perching birds—the order Passeriformes—containing over 55 percent of all bird species (although the voices of crows challenge the notion of singing!). These vocalizations are usually delivered from prominent perches, although larks and some other species may sing when flying and have evolved songs (and calls) somewhat in keeping with the structure of their habitat. Both cool and hot weather and rain and wind decrease the amount of singing.

A few species, such as red-eyed vireos, sing more or less all day long—a very repetitive song, leading to their previous common name, "preacher birds." Most birds vocalize more vigorously in the early morning and evening, with the amount of light, rather than the time of day, determining the beginning and end of their singing. Thus the dawn chorus may start at different times, but species will sing in the same order. A few vocalize at night, such as northern mockingbirds.

Most birds show a seasonal variation that is well correlated with breeding activities and hormone production. The richest, fullest songs invariably occur in the spring, when birds are establishing their territories and courting. After egg laying commences, they sing less, partly to avoid attracting predators. Should a male breed with a second female, in some instances because his mate has been killed, he resumes the full courting repertoire. After the breeding season, singing stops for many, unless they hold a winter territory.

The range of frequencies at which birds sing or call varies with the nature of the habitat and the ambient sound. The acoustic adaptation hypothesis predicts that narrow bandwidths and low frequencies, as well as long elements and interelement intervals, should be found in habitats with complex vegetation structures (which would absorb and muffle sounds), while high frequencies, broad bandwidths, high-frequency modulations (trills), and short elements and interelements may be expected in open habitats, without obstructive vegetation.

It has been hypothesized that the available frequency range is partitioned, so that an overlap of frequencies between different species is reduced. This idea has been termed an acoustic niche.

24 June | A Tidal Freshwater Wetland: The Most Beautiful Kind?

Although Jug Bay is halfway between the Patuxent River's headwaters and its entry into Chesapeake Bay, it is subject to the tidal rhythms of an estuary and the consequent movement and circulation of water. Its position favors the flow of the river and, thereby, freshwater conditions, with salinity, even during the dry summer months, rarely exceeding two parts per thousand. As the river's grade is low, the water flow does not undercut its banks, as it would in a streambed.

There are two high and low tides every 25 hours, with an average rise and fall of some 2.5 feet. Because the river flows through a somewhat narrow, flat flood-

Is Birdsong Music?

In *The Descent of Man*, first published in 1871, Darwin noted that male birds "charm the females by vocal or instrumental music of the most varied kinds," and Mozart, Vivaldi, and others composed music inspired by birds singing in cages or in nature. Scales, trills, themes and variations, and more are common to both, but for birds, should they be construed as music?

Marcelo Araya-Salas, a postdoctoral researcher at the Cornell Lab of Ornithology, studied the complex, musical-sounding song of tiny neotropical nightingale wrens and concluded that the resemblance between their song and music is nothing more than a coincidence. Of the 243 comparisons made between a nightingale wren's songs and musical scales, only six matched harmonic intervals (which occur when two notes are played at the same time). In his findings, Araya-Salas states that suggestions of a closer parallel between music and birdsong are misplaced and that "documented musical properties in birds might be caused by cultural biases of the listener or misunderstanding of the physics of musical compositions."

Not surprisingly, many scholars reject this view, one of the most compelling arguments coming from Emily Doolittle and Henrick Brumm in their study of the aptly named musician wren. These authors draw parallels between our music and the songs of wrens and other songbirds. The wrens favor consonant intervals (two notes having an overtone in common) over dissonant intervals (two notes considered to be somewhat unpleasant or tension-producing), which are heard in many human cultures. This species' musicality, however, goes even further, preferentially

plain, wind strength and water volume often magnify the tidal amplitude. Indeed, southerly winds in May could push tides two feet beyond the norm. Conversely, a northerly wind during a falling September tide can turn the marsh into a mudflat for some 12 hours and attract shorebirds.

The large volume of sediment entering the river upstream gives the water a muddy appearance. Consequently, submerged aquatic vegetation suffers from an absence of sunlight, resulting in a decline in habitat and food sources for wildlife. A further deterioration in water quality occurs in summertime, when the water

producing successive perfect octaves, fifths, and fourths, such that their songs sound musical to human listeners. The researchers even found passages with a striking similarity to Bach and Haydn. Nonetheless, as they note: "This does not mean that the musician wren is singing in a key the way a human musician might. Rather, the bird's preference for consonances leads to occasional conjunctions of pitches which sound to human listeners like they are drawn from the same scale."

Doolittle and Brumm asked 91 humans to compare successions of intervals taken from the musician wrens' songs with computer-generated songs in which the contours and durations remained the same but the intervals were slightly altered. The results were clear: human listeners considered the birds' interval choices to be more "musical."

They further caution that their results do not mean birdsong in general is constructed like human music. Each bird has its own way of singing, and some are not very musical at all.

It remains a mystery whether and how musician wrens discern musical intervals and how they structure their songs. The perception of intervals and other aspects of human music by nonhuman animals are of considerable relevance regarding the origin of human music and call for further investigation. Meanwhile, we can surely celebrate the "music" of both.

P.S. The musicologist F. Schuyler Mathews had no difficulty in declaring that bird song is music, and philosopher Charles Hartshorne suggests that "birds are primitive musicians, not advanced or sophisticated," since "every simple musical device, even transposition and simultaneous harmony, occurs in bird music."

column is enriched by nutrients and produces excessive amounts of algae (eutrophication).

Wetlands along significant portions of the river are effectively a transition zone between aquatic and terrestrial ecosystems, making them inherently different from each other, yet highly interdependent. The soils are either permanently or seasonally saturated with water; organic matter accumulates, as the lack of oxygen inhibits decomposition; and characteristic aquatic vegetation predominates, its composition varying with flooding.

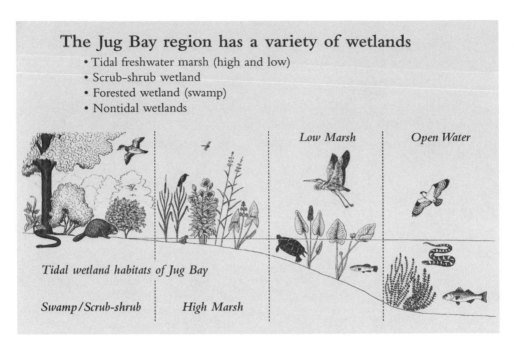

The Jug Bay region has a variety of wetlands
- Tidal freshwater marsh (high and low)
- Scrub–shrub wetland
- Forested wetland (swamp)
- Nontidal wetlands

Low Marsh   *Open Water*

Tidal wetland habitats of Jug Bay

Swamp/Scrub-shrub   *High Marsh*

The four types of wetland in the Jug Bay region. *Illustration by John Norton*

At Jug Bay, four types of wetland may be distinguished: high and low tidal freshwater marsh, scrub-shrub swamp, forested swamp, and nontidal wetlands. Forested wetlands border tidal freshwater marshes and are distinguished by the majority of woody plants growing in normally saturated or flooded areas. Trees there include red maples, green ashes, pumpkin ashes, sweetbay magnolias, and smooth alders (or hazel alders), as well as herbaceous species such as jewelweeds, halberdleaf tearthumbs, and lizard's tails. Nontidal wetlands exist farther away from the river's edge, bordering permanent and intermittent creeks flowing from the uplands. Scrub-shrub wetlands are even farther in but are still influenced by the tide. Here are found common buttonbushes, black willows, hazel alders, swamp roses, and crimsoneyed rosemallows.

Such plants support productive, diverse communities of aquatic and terrestrial animals, including invertebrates, fish, birds, amphibians, reptiles, and mammals. These organisms, in turn, form complex food webs linking tidal freshwater wetlands to adjacent riparian and terrestrial ecosystems. Red-winged blackbirds feed in the marsh but use scrub-shrub habitat for cover or as an occasional food source; North American beavers build dams to create nontidal ponds but often feed in its tidal portions; wood ducks nest in trees in swamps but feed in the marsh or in open water; and yellow perch occur in a variety of aquatic habitats.

Water depth affects the distribution of most species, and species assemblages in the marsh can change throughout the day with the tidal fluctuations. When the tide is high, fish and aquatic invertebrates, as well as some turtles and wading and swimming birds, forage in the standing water. On a falling tide, many organisms retreat to permanently flooded channels, while others bury themselves in the mud or occupy small pools. At very low tide, mudflats are exposed, allowing birds to search for invertebrates. In recent years, mudflats have been exposed less frequently, and fewer shorebirds—such as dowitchers, plovers, and spotted sandpipers—visit Jug Bay. In the air, swallows, flycatchers, bats, and other insectivores seek out varying abundances of prey, depending on the tidal stage and the winds.

Invertebrate fauna is poor in species, compared with nontidal zones. Macroinvertebrate communities are composed largely of worms in the sediments, and dipteran (fly) larvae and detritus-based processes involving bacteria and single-celled protozoa dominate the food web. Benthic organisms and the seeds of macrophytes (aquatic plants growing in or near water) are important links in the food chains leading to more-complex consumers. Major microbenthic organisms include amoebas and small, slender worms known as nematodes. Adult insects, especially beetles, flies, and dragon- and damselflies, are abundant on emergent vegetation, and some species may swarm high over the wetlands.

Of the vertebrate fauna, Jug Bay supports a mix of 46 freshwater, estuarine, and migratory species. Tessellated darters, largemouth bass, pumpkinseed sunfish, and redfin pickerels have an affinity for freshwater. There are some 9 species of amphibians and 16 reptiles using the tidal freshwater marsh. Of the former, freshwater forms are not generally common, largely because of fluctuating water levels, making their eggs and larvae easy prey for fish. Amphibian exceptions exist, however, in the regular appearance of green frogs, green treefrogs, spring peepers, Fowler's toads, and eastern American toads along the upper edges of the marsh. Aquatic turtle species and common watersnakes are the most abundant reptiles.

Compared with tidal wetlands, tidal freshwater wetlands support especially large numbers of birds and a high diversity of avian species. At Jug Bay, some

The Impact of Tides and Seasonal Changes

Tidal freshwater marshes on the East Coast of the United States have highly diverse plant life, with some 25 to 40 species growing in the tidal zone and over 100 species in marshes that are flooded infrequently. Studies at Jug Bay by Christopher Swarth, Patricia Delgado, and Dennis Whigham have recorded only 2–4 species growing in the low marsh, with spatterdock being dominant; 6–14 in the midlevel marsh; and 16–28 in the high marsh. This range of diversity indicates that in stressful low marsh areas with near constant inundation, far fewer species are able to survive, in contrast to slightly drier areas where many more species are found.

Seasonal changes are quite apparent, as well as those occurring through the growing season. Many tidal wetland species at Jug Bay die back during the winter, presenting the wetland with a flat, barren aspect, since the amount of decomposing plant material declines as spring approaches. By March to April, the high marsh greens up as annual species (e.g., jewelweed, halberdleaf tearthumb) respond to warming conditions. Canopy development in the high marsh sees the leaves and shoots of perennials (sweetflag, wild rice, green arrow arum, reed canarygrass) grow more rapidly than the seedlings of annuals, reaching a maximum height and biomass to form a canopy in early summer. By mid- to late summer, the stems of many species emerge above the perennials to form a canopy that is often over

142 species have been recorded. Seeds, insects, and small fish are abundant for dabbling waterfowl, soras, shorebirds, and songbirds. Microhabitats for foraging include the standing stalks of phragmites and cattails used by marsh wrens, sparrows, and blackbirds. The shrubs and trees at higher levels are used for foraging and nesting by songbirds and woodpeckers, and dead trees serve as perches and nesting sites for hawks, ospreys, and bald eagles. The ebb and flow of tides govern the composition of foraging species: shorebirds explore exposed mud; ospreys and herons seek areas during rising and falling tides to capture concentrations of fish; and grebes, ducks, gulls, and belted kingfishers find foraging spots at all stages of the tide. Waterbird abundance and diversity varies seasonally, with the greatest use of the tidal freshwater marsh occurring during the nonbreeding season, from

three feet tall, with some species exceeding six feet (tidalmarsh amaranths, great ragweeds, wild rice). Some perennials die back as they become overtopped by growing annuals taking advantage of the thinning of neighboring vegetation before reaching a maximum biomass in early autumn. The annuals then set seed and senesce (wither), thereby allowing light to penetrate closer to the soil surface and permitting a second growth period for perennials, such as green arrow arums. Swamp areas display similar seasonal changes, with herbaceous species starting to grow before the leaves of trees and shrubs emerge.

Seasonal changes in the canopy and in species that emerge produce a palate of colors. Most early flowering plant species have green flowers (e.g., sweetflag, green arrow arum, reed canarygrass), but with seasonal progression, there are impressive displays: the pink or white of crimsoneyed rosemallows, the magenta of purple loosestrife, the orange of jewelweeds, and the light yellow of wild rice. In the autumn, the tidal wetland becomes studded with the yellow flowers of the larger burr marigolds and common sneezeweeds, or of purplestem asters. In swampy areas, the senescing red leaves of maples and blackgums, the yellows of alders and ashes, and the purples of American sweetgums add their own palette. During winter, following the first frosts, the high marsh turns brown, offset somewhat by the lingering red fruit of common winterberries.

November through April. Generally, bird nesting is limited to midlevel and upper intertidal areas consisting of cattails, phragmites, purple loosestrife, and crimson-eyed rosemallows, as well as shrub zones of woody plants.

Mammals found in Jug Bay's tidal freshwater wetland are aquatic species and those of the neighboring upland forest. Muskrats are the most conspicuous, especially in cattails, green arrow arums, and wild rice habitats, where they build mound-like lodges, as well as burrows and runways. A variety of animals use the lodges for nesting and overwintering. Mallards and Canada geese nest there, and great blue herons, northern harriers, ospreys, and bald eagles often perch on them. North American beaver lodges are generally located in freshwater areas of Jug Bay, damming a tidal creek, although the beavers position lodges in many areas without building dams. North American river otters, American minks, and raccoons are top predators, and marsh rice rats are omnivores, feeding on crustaceans, insects, and plant matter.

Over time, the marshes of Jug Bay will become shallower as they accumulate sediments from the Patuxent River. Countering this effect are erosion and the gradual rise in sea level. Plants respond with high marsh species becoming more dominant along the shore.

25 June | EMPTY EGGSHELLS AND EGG SHAPE

Eggshells are often found scattered around the landscape at this peak of the breeding season, and one wonders if the chicks have hatched or been predated. Many females often eat the eggshells, to replace the calcium used to form them. Grebes thrust their eggshells underwater, releasing them far from the nest. Other avian species fly off to dispose of them. Molly Schumer and Jennifer Jin have shown that birds will preferentially remove white shells placed in their nests, as they compromise nest concealment. They generally fail to remove green shells, however, since it is a color that increases nest camouflage. Indeed, if the birds removed green objects around their nests, they might begin removing leaves and compromise the nest.

A behavior similar to eggshell removal has evolved in many birds: fecal sac removal. Many observers have found that eggs close to fecal residue suffer greater predation than those farther away. Both eggshell- and fecal sac–removing behaviors involve ridding the nest of predator-attracting elements, ultimately helping the offspring survive better.

Surprisingly, few eggs are egg shaped. Owl eggs are almost spherical, hummingbird eggs are elliptical but not asymmetric, and sandpiper eggs are almost teardrop shaped. Why this range of shapes, especially when eggs have the similar job of protecting and nourishing the chick? Some claim that they are asymmetric to prevent them rolling too far from the nest; others suggest that an elliptical shape gives them strength in the vertical axis.

An alternative answer suggests that it depends upon how fast the mother flies. Mary Stoddard and her colleagues offer an overarching rule: the more a bird's wing is designed for speed, the more streamlined the egg. They state that this is not because the egg needs to be elongated, but because, when a bird has to fly fast, its abdominal cavity, including the opening through which the egg passes, is slimmed down, forcing the egg's shape to change.

29 June | Butterflies Butterflying

What could be better on a sunny morning than wandering in a flower-filled meadow, looking up at the blue sky, breathing in the sweet scent of buttercups and hot grass, and delighting the eyes with butterflies dancing in search of a mate or simply basking in the sunlight. There are cloudless and clouded sulphurs, displaying bright yellow wings; the orange-to-brown-colored dun skippers and zabulon skippers; big, bold, beautiful brownish-orange great spangled fritillaries; and smaller pearl crescents. For a contrast, I admire spring azures, their pale silver-blue wings spotted with pale ivory dots. A common buckeye flies into view, with its target-shaped eyespots, white bands, and two prominent orange bars. Before me is pointillism, displayed on a huge canvas.

Later, in the marsh, eastern tiger swallowtails nectaring on swamp milkweed ignite my interest with their large, yellow-and-black tiger-striped wings and distinctive tails. The outer edges of their forewings and hindwings are black, with a row of yellow spots, while the inner margin of the hindwings has small red and blue spots. Males seek females by constantly patrolling habitats containing host plants for their larvae.[1]

Like humans, butterflies have courting styles and strategies as varied as they are ingenious. Research by Johann Anderson, Anna-Karin Borg-Karlsson, and Christer Wiklund has shown that when male cabbage whites, mustard whites, and orange-tips mate, they inject methyl salicylate along with their sperm. The smell of this compound repels other males and ensures the first male's paternity for the eggs. In contrast, a virgin female displaying a very similar posture releases a different chemical that will prolong the courtship ritual. Males are sensitive to these chemical and postural differences, and they can discriminate between a receptive virgin female and an unreceptive mated female. During swallowtail courtship, the sexes fly about each other, with the male releasing perfume-like chemicals that trigger a response in members of the same species (pheromones) to entice the female into mating.

30 June | DROUGHT?

May was green and lush, a product of abundant rain, while June received less than half the normal rainfall. The yearly deficit is almost four inches and a drought looms, as it did at April's end. Forecasters say that the past 12 months rank as the sixth driest on record, dating back to 1872, and note that a very dry June bodes problems for the rest of the summer.

I witness effects of the drought on yellow leaves, and those fallen to the ground crisp to my tread. Meadows and pastures are increasingly brown, and homeowners are much a'sprinkling their lawns. Streams run low, and tidal reaches edge upriver.

Forecasters record that this summer has been a jumble of cooler and warmer intervals, being warmer overall than normal. They predict that the pattern will continue into July.

"Moan of doves in immemorial elms and the murmur of innumerable bees," as Alfred, Lord Tennyson, describes June in "Come Down, O Maid." Drowsy insect music, grasses stroked by carefree winds, flutters of the first fledglings, and the gentle twittering of swallows now take center stage at this, the year's summit.

June marks the high point in nature's calendar. Warm temperatures stoke nature's furnace, and luxuriance in every dimension seduces the eye. Long days allow feeding the young to reach a frenzy. Birdsong has reached an apogee, although next month it will drain away, to be replaced by the cadences of insects.

A view of the Beaver Pond. *Rob McEachern*

JULY

The Heat Moon

The Summer looks out from her brazen tower,
Through the flashing bars of July.

Francis Thompson, "A Corymbus for Autumn"

1 July | A'BLOOMIN'

Summer-blooming plants now dominate the marsh. In the high marsh and the scrub-shrub zone, few rival crimsoneyed rosemallows, with their large rose-like flowers, ranging from pink to crimson, with dark red centers, and their densely compacted yellow stamens.[1] Bumblebees and other long-tongued bees now pay frequent visits to these flowers, and I have seen ruby-throated hummingbirds seek their nectar in earlier years. In shaded areas, lizard's tails, or water-dragons, abound, their white flowers packed on slender, tapering, stalked spikes assuming an arching, tail-like shape. As the greenish seeds develop, the "tail" assumes a wrinkled appearance; hence the common name.

Later, I find common buttonbushes—multistemmed shrubs with glossy dark leaves. Their small flowers are borne in distinctive spherical clusters, with a fringe of seed- or fruit-producing parts protruding beyond the white petals, taking on a distinctive pincushion-like appearance. The flower heads are very attractive to bees and butterflies, because of the copious nectar, and they will mature into hard, ball-like fruits (nutlets). In winter, these will be eaten by waterfowl.

The swollen joints, or nodes, of smartweed species are diagnostic, and I examine a tangled thicket of halberdleaf tearthumbs, a name derived from a brutal medieval weapon and the barbs on the plant's stem—all too capable of tearing the flesh! It has used these sharp hooks to climb up other plants, where its small pinkish-white flowers are visible at the end of a long stem. This species, along with a host of other smartweeds in the marsh—waterpeppers (also known as marshpepper knotweeds), dotted smartweeds, and arrowleaf tearthumbs—are important foods for waterfowl, soras, and seed-eating songbirds (e.g., swamp sparrows, red-winged blackbirds) later in the year. They also provide protective cover for wetland wildlife.

I sit against a tree and gaze at the marsh. The beautiful bright yellow, ball-like flowers of spatterdocks float on the water or stand above it, their emergent stems and leaves being caressed by a cooling breeze. The less glamorous violet-blue flowers of pickerelweeds form large beds in the shallow water, and I see bumblebees, carpenter bees, and syrphid flies (hoverflies) deriving nectar and pollen from these blossoms. Colorful dragonflies and damselflies dash over the marsh, and I hear the calls of ospreys, now feeding their young.

The English naturalist, Rev. John Banister, first collected pickerelweeds in colonial Virginia, circa 1680. Sad to note, while he was out gathering various plant specimens, a woodsman in his party mistakenly shot and killed him.

Cattails—showing their long, erect, blade-like leaves and characteristic female sausage-like flower heads, with a male spike—now dominate many portions of the marsh. Pure stands form "forests" of either broad-leaved or narrow-leaved cattails, in some cases producing hybrids. The former species prefers deeper water and is generally less tolerant of disturbance. The flowers of both produce prolific downy seeds (an average of 220,000 seeds per spike), and the leaves (held erect by struts and partitions, akin to the interior of airplane wings) are a boon to animal life. There is evidence of sap-sucking aphids and the larvae of moths mining the leaf tissue; other insects eat the flowers and seed spikes. Though difficult to observe, in other places I have seen marsh wrens, red-winged blackbirds, and swamp sparrows suspend their nests amid the dense stalks and leaves. Muskrats use cattail leaves and stems to construct their dome-shaped homes. In winter, the starchy, underwater, creeping rhizomes of these plants also provide food for muskrats, along with North American beavers and, sometimes, Canada geese.

I have been told to look for sac spiders and been directed to the leaf tips of cattails. Here I find a few folded over and tied down with silk to form nurseries and coffins, thus named because the female deposits an egg sac in the fold and dies in situ to provide food for her young.

I saunter to the edge of the woodland and encounter clustered bright orange flowers atop the upright hairy stems of butterfly milkweeds. True to its name, this species has attracted a host of butterflies because of its color and plentiful production of nectar. The leaves will be a good food source for monarch butterfly caterpillars in the early autumn.

In an even more open area, bumblebees and carpenter bees explore the spectacular climbing purple passionflowers, bearing five lavender petals, five white sepals, and a crown, or corona, of numerous fringe-like segments arising from above the petals.[2] Just visible are the beginnings of fleshy, egg-shaped, edible fruits, called maypops (another name for the plant, referring to the loud popping sound made when the fruits are stepped on).

While I was taking my leisurely walk, a storm tracked east from Chesapeake Bay to Washington, DC, unleashing a burst of wind and rain and taking on the dimensions of the apocalypse. Photographs showed that the storm's leading edge assumed a menacing dome of low-hanging clouds of doom, known as shelf clouds. The Sanctuary was buffeted but remained unscathed.

3 July | NESTS AND NESTLINGS

The nest, in the fork of a bush, of an American goldfinch is exquisite—a neat, rounded bowl of plant fibers laced with catkins and wool, and lined with the downy filaments of seedheads (especially those from thistles and milkweeds) and caterpillar webbing. The edges of the nest are bound with spiders' webs, and in the bowl there are five pale blue eggs.

Turning to the tallest branches of a red oak, I see a bulky mass of twigs, some of them leafy. My binoculars allow a closer inspection, and I estimate it to be a Cooper's hawk's nest, 12–15 inches high and 20 inches in diameter. From reading, I know it has a cup-shaped depression in the middle, lined with bark flakes.

I keep to the shadows and in time see a juvenile Cooper's hawk glide through the canopy as a "flying cross" and land on the nest. Another juvenile joins it, and one utters a loud, grating "cak-cak-cak." Both are an impressive sight, with creamy underparts highlighting the narrow crisp streaks on their wings and tails. They are now "branchers," making frequent returns to the nest for food, rest, and roosting. As they become more proficient flyers, they will wander farther away, leaving the nest after one month. Neither parent was observed.

Perhaps it's the memories of childhood that draw me to swallows and their nests. On a sheltered beam of the Observation Blind, overlooking South Glebe

Marsh, a pair of barn swallows has built their half-circular cup-shaped nest. It is a wonderful construction of mud pellets mixed with plant fragments. A few feathers poke from the cup's rim, and in between are four nestlings, showing their wide, hungry mouths. They call to their parents, and I retire to a corner in the hope that they will ignore me and feed their young. At times the adults sweep through the blind's openings, but they never land. I make a retreat and hear the calls of the nestlings as a parent returns.[3]

But there are further delights in store. A pair of tree swallows zoom in and out of a nest box attached to the blind, and I hear the greetings of their young. At the end of the Railroad Bed Trail, eastern kingbirds have built their nest halfway up the arm of a small crane, used to hoist boats into the water at the end of the River Pier. Compared with that of the swallows, this nest is a bulky, ragged-looking affair of coarse stems, with rootlets hanging from their edges. The parents are quite nonchalant and continue feeding in the presence of people launching kayaks.

On the ground near the Wetlands Center, I see a fledging eastern phoebe being fed by its parents. The young call a long drawn-out "tree" to attract the adults. This may well be the second brood.

4 July | Minks, Muskrats, and Otters: Competition in the Marsh?

I'm lucky to come across two American mink kits on the Railroad Bed Trail, for this species is often nocturnal. They were probably born in April or early May and began hunting at eight weeks of age. Their skipping in and out of the water and parting the vegetation is infectious and draws the attention of a young couple out for an afternoon stroll. We look in vain for their mother, as kits normally remain with her until autumn, for protection against birds of prey, owls, foxes, and otters.

Later, I find myself pondering the differences between American minks, North American river otters, and muskrats, as well as the extent of their competition in and around the marsh. First I tease out their morphological dissimilarities. Minks have elongated bodies, small ears, semiwebbed feet, and waterproof fur. In contrast, otters have large cylindrical bodies, rounded ears that barely project above their fur, a tail serving as a rudder, fully webbed feet, good vision in or out of the water, and enough lung capacity to forage underwater. Muskrats, although related to voles, are more beaver-like in their way of life. Their sleek bodies are covered in short, thick fur, and they have semiwebbed feet and unusually flattened tails, enabling them to be excellent swimmers.

Minks forage along the shoreline at night, otters always forage in the water at any time of the day, and muskrats are most active at night or near dawn and dusk, searching for aquatic vegetation. Food preferences are also quite different. Minks are generalist predators, consuming small fish, crayfish, amphibians, birds and their eggs, and small mammals. An otter's diet is almost exclusively of fish, invariably six inches long. Muskrats feed largely on lush vegetation, such as cattails and spatterdocks, and are capable of eliminating the plants in a wetland habitat. Not surprisingly, minks and otters belong to the taxonomic order Carnivora, whereas muskrats belong to the order Rodentia.

One may conclude with confidence that the niche separation of these mammals is quite marked: competition between the species occurs at the habitat margins, their feeding is restricted to certain areas and times of day, and their prey species are quite different.

5–6 and 9 July | A "Wallage" of Rain

Half the molecules in a raindrop may be older than the sun.

Casper Henderson, *A New Map of Wonders*

To Cornishmen, very rainy days produce a "wallage" of water, and a wallage it has been these past two days. Flooding has been reported in many parts of Maryland, and the Patuxent River has become a darker chocolate brown. The woodland streams gushed, vernal pools expanded, and vegetation was bent askew. Animals retired to safe places, and those venturing out took on a ragged appearance.

Usually raccoons, foxes, mice, rabbits, and other small mammals stay in their dens, constructed so they will not flood. Insects hide under leaves, using them as umbrellas. To keep dry, birds dip their bills into the oil glands near their tails and rub the oil over their feathers, thereby donning "raincoats." White-tailed deer go about their business in light rain, seeking shelter only during heavy precipitation.

The total amount of annual rainfall thus far barely makes 17 inches, against a norm of nearly 24 inches, and I do see leaves falling from some trees. Does this presage an early autumn?

10 July | EASTERN BOX TURTLES

A turtle draws the attention of a young couple as they wander along the Rail-road Bed Trail. A closer look reveals an eastern box turtle, ambling through leaf litter and heading toward the wetland's edge. Its reddish eyes mean it's a male, and at this time of year he is beginning to seek a female. At any major disturbance, a hinge on its underbelly (plastron) will allow it to tightly seal its shell from most attackers, so we observe the turtle at a safe distance.

Studies initiated by Christopher Swarth, a former director of the Jug Bay Wet-lands Sanctuary, have identified nearly 600 individuals within a 200-acre area of the Sanctuary. Volunteers and Sanctuary staff record measurements on every eastern box turtle found and use mark-recapture techniques and radio telemetry to study their movements. This helps us to better understand their use of habitats throughout their annual cycles.

The work has focused on females, because their habitat needs are greater, since they require both foraging and nesting areas. The production of viable eggs depends on the female's access to nutritious and abundant food. Egg development, in turn, relies on a suitable nesting habitat away from the floodplain, to avoid being submerged.

Eastern box turtles typically lay their eggs during the first two weeks of June, placing four to six eggs in a nest made by digging a hole in the ground, covering them with dirt, and then abandoning them. With warming soil incubating the eggs, between 80 and 100 percent will hatch if mammalian predators do not discover them. Female box turtles begin to lay eggs when they reach sexual maturity, around age 5 to 10, and continue egg laying throughout their lifespan, which occasionally reaches 70 years or more.

Telemetry studies by Michael Marchand and colleagues allow the turtles' home range—where they spend their time over the course of a year—to be mapped. A home range includes all of the resources needed for survival, and aware-ness of this is critical to a resource manager when planning the conservation of a species. While a turtle's typical home range may cover four to five hectares, or 10–12 acres, in the first season of the study one female turtle's home range covered about five acres. During the second season, on two occasions trackers found her

An eastern box turtle. *Rob McEachern*

outside the sanctuary, possibly attempting to nest in adjacent farm fields. This stretched her home range to 25 acres.

Late fall is a signal for the turtles to settle in for a long winter's nap. Accordingly, they hibernate under a layer of leaf litter, either to the depth of their shells or an inch deeper, so that the top of the shell is an inch or two below the surface of the soil. In winter, the insulating properties of soil and leaves keep the subsoil environment from becoming as cold as the ambient temperature. Such conditions also keep a turtle cooler in spring, thus reducing the chance that it might emerge prematurely on unusually warm early spring days. A cold snap could actually kill a turtle if it was on the surface and unable to dig under quickly enough to avoid succumbing from freezing temperatures. Turtles therefore remain in this insulating blanket until midspring, when they emerge to roam and nest.

Though not endangered, eastern box turtles, like all turtle species in Maryland, live in ever-contracting habitats. Tracking has showed that box turtles use wetlands for significant amounts of time and, equally, that they need a mix of habitats: a forest for food and cool temperatures; wetlands for rehydration and food; and sunny, open sites for mating and nesting. They also require corridors to provide areas of safe passage between these habitats.

Corridors are important not only to turtles, but also to a variety of animals that migrate short or long distances during their life cycle. Housing developments, roads, and strip malls act as barriers in the corridors, cutting off wild creatures from critical habitat and from one another.

Loss of habitat, collisions with cars, and illegal trade have made eastern box turtles scarce on many of their ranges.

VERNAL POOLS

Chuck Hatcher tells me that with the exception of Mark's Pond, by mid-June all of the vernal pools were dried up, signaling the end of the breeding season for spotted salamanders. Those whose eggs did not hatch, or grow into mature forms, or make for land by late May or early June were unlikely to survive. If and when these pools fill up following summer rains, mosquitos, dragonflies, and other insects will rule the roost. Come fall, marbled salamanders will lay their eggs under logs or in the mud of this year's ponds and remain ready, once winter precipitation begins to fill them, to start their cycle anew.

Mark's Pond is truly a gem. It provides a deep pool that is less susceptible to the whims of nature's amounts of rainfall.

11 July | MAGICAL FIREFLIES AND WONDERFUL BATS

In late evening, male fireflies begin to dance over the meadow at River Farm. Hundreds of them, in the grass, are flashing light from their abdomens. Each species has a distinctive flash pattern to attract females. The latter wait before responding with a blink or two. For some males, this blink may prove to be a deadly ruse used by a larger species that has cracked its prey's flashing code, and they succumb to these femme fatales. Sometimes males have trouble finding mates in the grass, or females do not return the males' ardent flashes. But the pressure is on, for fireflies spend years underground as larvae, emerging to fly for just one to two weeks. As with all life forms, their mission is to reproduce and, at this, more than half of the males will fail.

Fireflies, or lightning bugs, are beetles. There are carnivorous and herbivorous ones, both encompassing a highly specialized four-stage life cycle. After mating, the female lays her eggs in moist soil, often near the edge of a stream or marsh. Upon hatching, the larvae (commonly called glowworms, because of a pair of small

The Newsletter

Many conservation groups devote time and energy to informing their members, who, in turn, pay homage to such diligence. The newsletter for the Sanctuary, first issued in March 1986 as copied typed sheets, has undergone many changes in format and content. Always lively and informative, it presents a descriptive account of the evolution of the Sanctuary's administration and management of the various habitats, as well as information about research and educational efforts.

It began as the *Jug Bay Sanctuary News*, with a summary of events held (the summer science camp) and a calendar of those to come (lectures, workshops, children's day, Friends of Jug Bay general meeting). It also presented an overview by the director of the Sanctuary: a hot and dry summer, with an early migration of waterfowl and soras, nesting turtles struggling to find moist conditions (one red-bellied turtle chose to dig at the bottom of a gutter spout), and the wilting of herbaceous plants. Later issues covered the marsh monitoring program (an inventory of amphibians and reptiles, birds, fish, insects, plants, and water quality) and ecological studies of various fauna and flora. Volunteers working with educational groups, research projects, the establishment of volunteer projects, and the creation of a naturalist's library were chronicled, as well as the development of the Friends of Jug Bay, which fosters educational and research programs and has created a system of trails.

Its current format, *Marsh Notes*, appeared as volume 5, number 2 of the Sanctuary's newsletter in summer 1990 and began to include articles on special topics. The following edition ran one on "Mummichogs: Mighty midgets of the marsh."

Now in its thirty-first year, *Marsh Notes* maintains the high standard of informing members about research initiatives, giving profiles of species and their habitats, and describing conservation activities, education events, public programs, and service awards.

spots near the end of the abdomen, used to warn off predators), feed on worms, insects, mites, and snails.

As autumn progresses, the partly grown larvae seek shelter under the bark of trees, a log, or a tree stump and hibernate during the coming winter. They emerge in the spring to feed and pupate, and achieve adulthood during the summer, when they consume pollen or other plant materials.

I do not know how many firefly species weave their golden lights before me, but the color, the time lapse between flashes, and the intensity and length of the pulses suggest that there are many fireflies present. Each one faces challenges during its yearlong cycle, occupying different niches during its various life stages and having to face spiders and other predators, all in the hope of reproducing. When eaten, their tissues are recycled into forms large and small.

Before darkness falls, eastern red bats begin their nightly routine in the forest and along the marsh. They are fast fliers, pursuing moths, flies, true bugs, beetles, cicadas, and even ground-dwelling crickets that they snap up from the woodland floor. But they avoid fireflies, which are distasteful and can be poisonous to vertebrate predators. Sometimes the bats take steep dives, pulling away from the ground with only inches to spare. On other occasions, they swoop down for a drink from the surface of the marsh. Time and again, I marvel at these denizens of the night as they flow through their aerial hunts.

During the day, the bats have been roosting in the dense foliage of deciduous trees, where their coloration blends with the foliage. When they hang upside down, they look like dead leaves or, if they choose conifers, pine cones.

Females gave birth to one to four pups during late May and early June, and I am looking at both juveniles and adults swooping low and then high, their silent flight broken only by the calls of katydids and crickets. Mating occurs in late summer or autumn, with the sperm stored in the female's reproductive tract until spring, when ovulation and fertilization take place.

From September to November, eastern red bats move south to avoid severe winter temperatures, travelling north again between March and April.

There are currently no reports of mortalities for this species as a result of white-nose syndrome. It's a disease caused by a cold-loving fungus, which afflicts bats hibernating in caves and mines and continues to expand its range in North America.

12 July | THE ESSENCE OF SUMMER

It's been in the high 90s these last few days, with matching humidity. With a heat index of over 100°F soon predicted to come, I set off for Jug Bay before the temperatures become intolerable and confine me indoors.

I find the marsh shimmering in the early morning heat. Spatterdocks, with their broad leaves, as well as green arrow arums and pickerelweeds in the low marsh next to the open water, all sway in the trifling breeze, their flowers attracting butterflies and bees. With their differing hues of green, stands of wild rice, cattails, and phragmites have now reached 10 feet in height and dominate all but the meandering river and its creeks. By early September, the needle-like seeds of wild rice will have matured enough to fuel the long flights of migrants, especially soras.

Ospreys peruse the open water in search of a catch for their young, which are now nearing their test flight stage. At the water's edge, a great blue heron is poised to strike, and a wood duck flies upstream, quickly disappearing, with angled elegance, into the trees bordering the marsh. Birdsong is becoming muted, and the air is now dancing with buzzing bees, wasps, and flies, as well as ringing with the incessant drumming of cicadas.

In the shade of smooth alders and red maples, I follow the Railroad Bed Trail dividing Upper and Lower Glebe Marshes and notice that the bordering flowers attract diminutive eastern long-tailed skipper and least skipper butterflies. I tread carefully to avoid crunching coal residue spewed from bygone trains, but I have alerted the next object of my interest in the marsh: adult and young red-winged blackbirds clutching the stalks of cattails and phragmites. It is a noisy spectacle as the young clamor for food from their parents. Meanwhile, a green heron whooshed directly over my head, and I felt the push of air from its beating wings. Nearby, a family of common yellowthroats has been searching for insects on swamp rosemallows and jewelweeds, whose delicate flowers luxuriate in the morning sun. All slowly slink away.

I am still in the marsh, and high up on a bare branch of a smooth alder, three fledgling tree swallows await food from their parents. There's much commotion when the adults arrive, who seem to distribute the nourishment evenly. One of the fledglings takes off to follow a parent "hawking" insects.[4] The two turn and meet chest to chest, the adult deftly transferring its catch. Both then peel away to zoom over the marsh, the fledgling joining its reflection as it breaks the river's surface with its beak to scoop up a drink.

From the River Pier at the end of the Railroad Bed Trail, a double-crested cormorant perches on the cylindrical base of a former trestle, now removed, that had allowed an occasional steamboat to pass. It grows fearful of an approaching boat and takes off down river. An otter glides upstream, taking advantage of the incoming tide, then drifts to the river's edge and disappears into the reeds overlooked by Mount Calvert.

I return to higher ground in time to hear the throaty call of a yellow-billed cuckoo and the delicate "fee-be-o" of an alder flycatcher. A search for both birds proves to be fruitless, but I accidently disturb an adult wood duck. It cries out as it and its three young rapidly leave their nest in the hollow of a tree and fly with impressive speed over the marsh.

It's time to sit, and I settle on the mossy knees of an American beech. Apart from the incessant song of a red-eyed vireo (living up to its other common name, preacher bird), I hear only the haunting song of a wood thrush competing with the concerted chatter of insects. On a summer afternoon such as this, its call must be one of the most exquisite and evocative sounds in all nature. Indeed, in the 5 July 1852 entry in his *Journals*, Henry David Thoreau observed that "the thrush alone declares the immortal wealth and vigor that is in the forest," adding that "whenever a man hears it, he is young, and Nature is in her spring."

At the same spot in late evening, butterflies, cicadas, and flies move between woodland and meadow, and dragonflies flit between marsh and meadow, their lives dependent on habitat diversity. Later still, mosquitoes and swarms of various species of flies leave the forest and fill the air above the meadow, many to be grazed on by bats and dragonflies. All is focused on one aim: to find nourishment and reproduce before the coming winter.

At the Wetlands Center, I learn that a juvenile lesser yellowlegs was seen earlier in the week. This timing is quite early for its migration south and is ahead of millions of other birds shortly to begin their fall flight to wintering and stopover habitats.

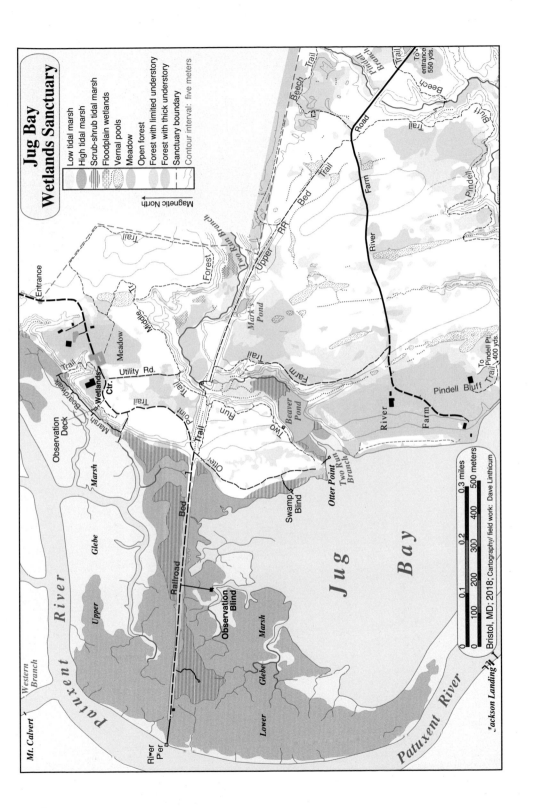

Jug Bay Wetlands Sanctuary

Low tidal marsh
High tidal marsh
Scrub-shrub tidal marsh
Floodplain wetlands
Vernal pools
Meadow
Open forest
Forest with limited understory
Forest with thick understory
Sanctuary boundary
Contour interval: five meters

Magnetic North

Mt. Calvert

Western Branch

Patuxent River

Upper Glebe Marsh

Lower Glebe Marsh

Observation Blind

Jug Bay

Patuxent River

Jackson Landing

River Per

Railroad Bed

Observation Deck

Boardwalk

Marsh Trail

Entrance

Wetlands Ctr.

Utility Rd.

Meadow

Middle

Forest Tr.

Trail

Point Trail

Otter Trail

Two Run Branch

Swamp Blind

Otter Point

Two Run Branch

Beaver Pond

Farm Trail

Mark's Pond

Upper RR

Bed Trail

River

Farm Road

Beech Trail

Pindell Branch

Beech Trail

Pindell Bluff

Trail

Bluff

Pindell

River

Farm

To Pindell Pt. Trail, 400 yds.

To entrance 550 yds.

0   0.1   0.2   0.3 miles
0   100 200 300 400 500 meters

Bristol, MD; 2018; Cartography/ field work: Dave Linthicum

PLATE 1. A partridgeberry, with its fruit. *Jim Brighton*

PLATE 2. Emergent skunk cabbage flowers. *Jared Satchell*

PLATE 3. A white-breasted nuthatch, one of the "Gang of Four."
Frode Jacobsen

PLATE 4. Skunk cabbages rapidly expanding in a creek
entering Upper Glebe Marsh. Bill Hubick

PLATE 5. A spotted salamander. Kerry Wixted

Plate 6. Eastern redbud flowers. *Bill Hubick*

Plate 7. A common yellowthroat. *Frode Jacobsen*

PLATE 8. A male wood duck. *M-NCPPC Department of Parks and Recreation*

PLATE 9. Northern watersnakes. *Frode Jacobsen*

PLATE 10. An American bullfrog. *Frode Jacobsen*

Plate 11. A prothonotary warbler. *Frode Jacobsen*

Plate 12. An adult 17-year cicada. *Richard Orr*

Plate 13. An adult blue dasher dragonfly. *Frode Jacobsen*

PLATE 14. Six northern red-bellied turtles warming on a log in the Beaver Pond. *Rob McEachern*

PLATE 15. A butterfly afternoon, with a common buckeye. *Frode Jacobsen*

PLATE 16. An eastern tiger swallowtail. *Frode Jacobsen*

PLATES 17. Marsh splendor, with pickerelweeds (*right*) and a close-up of one of the
blossoms (*left*). *Jug Bay Wetlands Sanctuary*

PLATE 18. A halberdleaf rosemallow. *Jug Bay
Wetlands Sanctuary*

PLATE 19. A tree swallow preparing to enter a nest box.
Rob McEachern

PLATE 20. A North American river otter eating a fish. *Rob McEachern*

PLATE 21. Three young osprey nestlings, with a parent.
M-NCPPC Department of Parks and Recreation

PLATE 22. A citrine forktail damselfly. *Frode Jacobsen*

PLATE 23. A hummingbird clearwing moth extracting nectar
from a spotted knapweed. *Frode Jacobsen*

Plate 24. A cluster of turbulent phosphila moth caterpillars.
Gary Van Velsir

Plate 25. A spotted turtle. *Rob McEachern*

PLATE 26. A male common whitetail dragonfly. *Frode Jacobsen*

PLATE 27. A monarch butterfly. *Frode Jacobsen*

PLATE 28. A marsh rice rat. *James Parnell*

PLATE 29. The Jug Bay Wetlands Sanctuary's marsh, with golden bands of burr marigolds, beige cattails, and dying spatterdocks (*foreground*). *M-NCPPC Department of Parks and Recreation*

PLATE 30. An eastern chipmunk. *Annette Allor*

PLATE 31. A male Cooper's hawk in flight. *Frode Jacobsen*

PLATE 32. A winter landscape in the Jug Bay Wetlands Sanctuary.
M-NCPPC Department of Parks and Recreation

14 July | Ospreys

On a hot and sultry morning, we set off down the Patuxent River to inspect osprey nests and band any occupying fledglings. Greg Kearns, the Patuxent River Park's naturalist, has been monitoring ospreys for more than 40 years and has gathered his student interns, who are members of the Maryland Conservation Corps, and visitors to participate in his efforts.[5]

As we leave the dock, well above us an eastern kingbird attacks an adult bald eagle attempting to cross the river. It dives on the eagle several times and then, to our amazement, clutches the eagle's head. Though the kingbird is eventually dislodged, it continues attacking until the eagle is well past midriver. The kingbird climbs higher and, seeming to celebrate its success, glides down, tumbling in midair (in what may be described as "tumble flight") and ending with a fluttering sashay before reaching a perch.

At our first stop, we ease the boat through pickerelweeds and carefully approach a nest platform. An adult osprey takes off, and its fledgling crouches in a flattish collection of sticks and branches. Anticipating that it, too, will join its parent, Greg carefully stretches out to seize the fledgling by the legs and secure it for banding. A member of the conservation corps holds the bird while it is banded and details are recorded about its sex (female), weight (three pounds), and age (seven weeks). She is a very healthy youngster. We take a final look before Greg's deft handling places her back on the nest. We retire and see the adult quickly return. At eight weeks, its offspring will fly.

Records tell us that there were three chicks in the nest a few weeks ago. It's possible that great horned owls ate the young, although great blue herons or crows may equally be the culprits. Overall, there has been a 50 percent breeding success in the Jug Bay area, although sites downriver have seen a dramatic fall, to 20 percent. Cold northeasterly winds, rain, and water temperatures barely in the 50s in May led to a two-week delay in egg laying and subsequent adverse impacts. Some females may have become agitated and left their nests.

At one site, we inspect a wood duck nest box below an osprey platform. We count 13 egg membranes plus one rotten egg and assume 13 hatchings have occurred. According to Greg, 800 eggs have been laid by wood ducks in the Jug Bay area, making it one of the season's best results.

An osprey reaching for prey. *M-NCPPC Department of Parks and Recreation*

Satiated with wonder and delight, we disembark and immediately spot a green treefrog sitting in the middle of the deck. It's a slippery customer with its remarkable jumps and lithe body, and many attempts are made before someone catches it. The frog is two-and-a-half inches long, lime green flecked with gold, and embroidered with a stripe along its sides, in great contrast to a very pale abdomen. Normally these frogs are nocturnal creatures, so we are lucky to see one. This species has been an occupant of the area only since the 1980s, slowly tracking northward.

A severe thunderstorm sweeps through the area later, starting at 5:30 pm and lasting until 9:00 pm. It has brought extremely heavy rain and strong winds. The heat index has soared above 100°F, making 13 July the summer's hottest day.

20 July | NATURE'S FODDER AND A TWINGE OF AUTUMN

Milkweeds are erupting, with their highly fragrant, domed, slightly drooping clusters (umbels) of fragrant, pinkish-purple or orange-red flowers in the upper leaf axils (the angle between a leaf and a stem). They are being plundered by bumblebees, honeybees, swallowtails, monarchs, and other insects. I see yellow pollinia (masses of pollen grains) on the legs of some of the insects and recall occasions when I saw the tongues or feet of individuals trapped in pollinia pockets, their bodies dangling from the flowers. That's a sight to arrest anyone's attention on a summer's day.

There is some exclusivity to milkweeds and certain other plants, such as wreath goldenrods. Insects that have stored the toxins of milkweeds in their bodies display a bright warning coloration, making them often unpalatable or poisonous to birds. Mammalian herbivores don't eat this plant, because of the bitterness of the leaves and their toxic properties.

We shall witness the arrival of monarch butterflies later, as they lay their eggs to produce another generation on the wing before wintering south of the border.

I stand in the shade at the marsh's edge and am quite captivated by the blossoms of cardinalflowers. This flower has brilliant, vibrant red petals and, as naturalist John Burroughs wrote, "it is not so much something colored as it is color itself." The flowers are very attractive to butterflies and hummingbirds, but not to cardinals! Most insects struggle to navigate the long necks of the trumpet-shaped flowers, so the plant depends on hummingbirds for fertilization. They begin by foraging in the middle of the blossoming upright stem, methodically working their way upward. This male stage of the plant produces the most nectar, and the hummingbirds concentrate their efforts on the top half, where the flowers are youngest.

As a hummingbird hovers at the opening of the flower to sip the nectar, the top of its head brushes the flower's reproductive parts. Should the flower be in the male phase, the hummingbird's head will brush against the tuft of hairs at the end of the reproductive tube and receive a dusting of pollen. If the flower is in the female phase, pollen already on the hummingbird's head will stick to the stigmas, or female parts. Since male- and female-phase flowers are present at the same time, hummingbirds usually are successful in transferring pollen to female-phase flowers. Small bees, invariably denied direct access to the nectar because of the flower's tubular shape, find slits in the flower to gain entry.

I am hearing reports from the Maryland coast of migrant shorebirds returning from the north: yellowlegs, red knots, and ruddy turnstones in mid-July and sanderlings, least sandpipers, and semipalmated sandpipers during the past few days. There are also reports of migrant Tennessee, orange-crowned, Cape May, and palm warblers moving south. Normally, most of these do not appear until late August.

Biological Annihilation

A study released earlier in the month paints a grim portrait of wildlife around the planet. "This is the case of a biological annihilation occurring globally," said Stanford University professor Rodolfo Dirzo, one of the coauthors. It means that a predicted sixth extinction in Earth's history is underway and is more severe than previously feared.

Eschewing the normally sober tone of scientific papers, the study concludes that what it terms "biological annihilation" represents "a frightening assault on the foundations of human civilization" and blames overpopulation and overconsumption for the crisis. Wildlife is dying because of habitat destruction, overhunting, toxic pollution, the invasion of alien species, disease, and climate change.

While previous studies demonstrated that species are facing extinction at an increasing rate, they gave the impression that it was a gradual loss of biodiversity. The new study takes a broader view. It assesses many common species losing populations as their ranges shrink, but which remain present elsewhere. These scientists found that billions of populations of mammals, birds, reptiles, and amphibians have been lost all over the planet, leading them to say a sixth mass extinction has already progressed further than was thought.

Detailed data indicate that almost half of the land mammals lost 80 percent of their range in the past century. Tropical regions have seen the highest number of declining species. In South Asia and Southeast Asia, large-bodied mammals have lost more than four-fifths of their historical ranges. While fewer species are disappearing in temperate zones, the percentage is just as high or higher. As many as half of the number of animals that once shared our planet are no longer present, a

At Jug Bay, a postbreeding flock of great egrets grace the marsh, and flocks of European starlings and red-winged blackbirds gather and cavort in the phragmites and cattails. A blackgum tree has developed bright scarlet leaves, its autumn color. The creamy white flowers of devil's walkingsticks and a string of 5 to 10 purplish maroon to brown flowers of groundnuts now dot the woodland understory. In the marsh, wild rice is beginning to seed.

loss the authors describe as "a massive erosion of the greatest biological diversity in the history of Earth."

The study team cautions that beyond any moral imperative, there are practical reasons to regret the eclipse of animals, whether megafauna or smaller and less "charismatic" creatures. The vanishing of a top-level carnivore or herbivore can have a cascading effect on the food chain, disrupting entire ecosystems. Other species directly provide "services" to humans, such as honeybees that pollinate crops or birds that ensure pest control. Previous studies show that ecosystems under stress, while resilient, have a breaking point—rapid change can lead to collapse.

"The time to act is very short," says Paul Ehrlich, another coauthor for the study. "It will, sadly, take a long time to humanely begin the population shrinkage required if civilization is to survive, but much could be done on the consumption front and with 'band-aids'—wildlife reserves, diversity protection laws—in the meantime."

In Maryland, more than 600 species and subspecies of plants and animals are listed in state regulations as endangered, threatened, in need of conservation, or endangered extirpated. Most of the state-listed endangered species are plants (263), and 96 are animals. An additional 70 plants and 20 animals are listed as threatened, and 35 animals are listed as in need of conservation. According to the *Maryland State Wildlife Action Plan 2015–2025*, the 86 plants and 32 animals listed as endangered extirpated species are thought to be no longer present in the state.

We are in very real danger of holding a mirror to nature and seeing nothing but ourselves.

Sources: Ceballos, Ehrlich, and Dirzo 2017; Maryland Department of Natural Resources 2016.

24 July | WETTEST WEEKEND

With wave after wave of downpours, this weekend has become the area's wettest of the past two years and one of the hottest and muggiest. Storms have unleashed over two inches of rain, accompanied by extraordinary lightning, the blackest of clouds, and, finally, rainbows. A tornado has wrought considerable damage in nearby Queen Anne's County, with a top wind speed of 125 mph, and a waterspout developed over the water just south of the Chesapeake Bay Bridge. Tornados are not rare in the Washington, DC, region, but few are as strong.

The sodden woodland is singularly depressing, and the marsh has assumed a leaden aspect, merging water and sky. A bedraggled northern cardinal offers slurred, repeated whistles, and a few squirrels stir in the shaken trees. Rain pitter-pattering on leaves gives way to rivulets and bombards the woodland floor.

For all that has transpired, we are still short of the normal amount of precipitation.

25 July | MOTHS

A warm evening and a bright light will attract some of the 27 species of moths recorded at Jug Bay. My hope is that by dusk they will fall or tumble down (with a soft landing) into a trap I have placed in the woodland to observe them more easily. Nocturnal moths use the night sky for navigation, and the trap's bright light serves to confuse and direct them to the container. After my inspection, they are released to complete their brief lives.

The first candidate—a beautiful large nais tiger moth—zooms out of the dark and, after much fluttering, falls to earth only to rise briefly and become trapped. Its abdomen is light pink, with a dorsal row of black spots. The forewings are mostly pinkish white, with small black wedges. Next, out of nowhere, an American ermine moth arrives, and I see full justification for its name. It has brilliant white wings, with dark fleck marks, and possesses a fluffy thorax and abdomen. Southern flannel moths are also aptly named, with a body covered in long, dull orange fur, hairy legs, and fuzzy black feet. They are sometimes called woolly slugs.

Many moth species do not feed as an adult; their only task is reproduction. To find a mate, males possess feathered antennae that are acutely sensitive to pheromones borne on the night air. Female moths release approximately one billionth of a gram of pheromones from a special gland located on the abdomen. Males of the same species can detect it within 300 feet and fly toward the source to find a mate.

As Jeanne Grunert notes, Swedish scientists recently discovered that male moth pheromones communicate detailed information to potential mates: age, reproductive fitness, and ancestry. This indicates a more complex courtship model for moths than was originally thought. It's also consistent with Darwin's prediction, in *The Descent of Man*, that sexual selection favors exaggerated sensory receptor structures, like antennae.

The dusk has given way to darkness, and more moths (plus a few beetles) invite my attention. I am much taken with a crocus geometer moth, with its slender abdomen and broad, pale yellow wings giving it a butterfly-like appearance. Then comes a brown male gypsy moth, an invasive species that has caused widespread defoliation of the trees in the Northeast after spreading from Medford, Massachusetts, in 1868.[6] Though the white-colored females cannot fly, people transporting egg masses in firewood have been responsible for its spread.

As I leave the woodland, my thoughts turn to the larval (caterpillar) stages of these and other moths. They frequently assume bizarre forms, are often poisonous, and possess an incredible diversity of structural defenses and cryptic coloration. I am also reminded of the skinny, orange-colored ailanthus webworm moth that looks more like a beetle, thanks to its colorful pattern and tightly closed wings, and has an appreciation for the flowers of its namesake, *Ailanthus altissima* (tree of heaven). Along with many other moths, it is active in daylight and resembles a wasp in flight.

Strange as these forms may be, I am inclined to nominate the young larvae of eastern tiger swallowtails as the most inventive. The first instars (a larvae stage of insects just hatched from eggs) are dark, with a large white spot on the abdomen, and they resemble bird droppings. After molting to the fourth instar, these caterpillars become green, with a swollen thorax and a transverse band of faint blue dots on each abdominal segment. On the hindmost thorax, the larvae also have a single pair of false eyespots—yellow ringed with black—that contains a smaller blue spot lined with black and a black line behind the blue spot.

The eyespots have been described as an example of the "terrifying devices" nature has provided to frighten away enemies.

28 July | DAMSELFLIES AND BEYOND

Although damselflies are similar to dragonflies, they have smaller and slimmer bodies and generally fold their wings above the body. They do, however, share a marked sexual dimorphism (with males being more brightly colored), metamorphose through a nymphal stage,[7] and have elaborately choreographed courtship repertoires.

Both groups reproduce using indirect insemination, with a mating pair forming a familiar heart or wheel shape, with the male clasping the female at the back of the head, while the female curls her abdomen down to pick up sperm from secondary genitalia at the base of the male's abdomen. A pair often remains together, with the male still clasping the female, as she lays eggs within plant stems in or near water. Like dragonflies, the hatched aquatic larvae, or naiads, of damselflies come equipped with a specialized underlip that can spring out with lightning speed to grasp unsuspecting insect larvae, tadpoles, or small fish. The naiads grow and molt into flying immature adults, whose color has yet to be developed.

The damselflies I see hovering over the marsh and quickly darting away are predators on other insects, especially mosquitoes and midges. They include a familiar male common blue damselfly, with its bright blue and black bands and stripes, the yellow-and-black markings of a citrine forktail, and an ebony jewelwing, with a metallic blue-green body and black wings. Between flights, they settle on nearby green arrow arums and I can enjoy their coloration.

Both damselflies and dragonflies are a favorite prey of birds and fish. The digestive systems of red-winged blackbirds and swallows convert these exquisite packets of protein into the music of an evening.

While observing insect and bird life, I am drawn to fingernail clams, attached to the stems of green arrow arums or freely patrolling the water's muddy bottom. They are beige, thin-shelled mollusks that can attach themselves, as tiny juveniles, to the feet of birds. Sometimes they are ingested, only to be regurgitated alive some distance from their birthplace.

Darwin took a great interest in the phenomenon of how sedentary mollusks may be transported from one watershed to another. To test the notion that mollusks are attached to the feet of ducks, Darwin used an aquarium in which he dangled a duck's feet and wrote, in *The Origin of Species*, first published in 1859: "These just hatched molluscs, though aquatic in their nature, survived on the duck's feet, in damp air, from twelve to twenty hours; and in this length of time a duck or heron might fly at least six or seven hundred miles, and would be sure to alight on a pool or rivulet, if blown across the sea to an oceanic island or any other distant point." This chance dispersal view championed by Darwin would develop into the science of biogeography.

28–29 *July* | Winter Nor'easter

An unusual storm system has hit the Washington, DC, area, resembling a winter nor'easter but with more-humid air generating significantly more rain. A strong jet stream aloft and a coastal low-pressure system stalling along the Delmarva Peninsula have brought extremely heavy downpours and chilly temperatures for late July, dipping into the 60s, with breezy winds from the north and northwest. Total rainfall has varied from two to four inches, in some cases the most rain falling in a single day for two years. For July, the data suggest that the storm is a 1-in-50-year event.

31 *July* | Aqualungs

The many ways in which insects have invaded and adapted to the world's myriad habitats is truly astonishing. Mountaintops, frozen wastes, and deserts have all been conquered. But as I amble along the marsh boardwalk, it is those living atop and under the water that intrigue me. For here are water striders (pond skaters), giant diving beetles, and whirligig beetles.

The striders are true bugs and have a unique ability to walk on water, including in some marine environments. Looking closely, I can see how they use the water's high surface tension to their advantage. Their long, flexible, slender legs distribute their weight over a large area, and water-resistant hairs lining the body prevent them from getting wet. But these hairs do even more. If a water strider were to be submerged accidently, the tiny hairs trap air, buoying the strider to the surface again while allowing it to breathe underwater.

For movement, the strider's middle legs have particularly well-developed fringe hairs on the tibia and tarsus to help increase their ability to row, with the hind pair being used for steering. When rowing begins, the middle tarsi are quickly pressed down and backward, creating a circular surface wave in which the crest is used to propel the insect forward. As a result, water striders often move at one meter per second or faster.

Diving beetles carry an air bubble underneath the abdomen (in an elytra cavity), enabling them to remain submerged for long periods. Though the bubble acts like a gill, allowing the beetle to extract oxygen directly from the water, the oxygen

A water strider. *Kerry Wixted*

supply must be replenished from time to time by breaking the surface tension of the water with its abdomen.[8]

The hefty bite imparted by oval giant water beetles, or giant water bugs, is often more than a match for dragonfly nymphs, crustaceans, tadpoles, and, sometimes, small frogs. These beetles can reach three inches in length. The females of some species cement 100 or more eggs to a male's back, with the eggs hatching after one week.

Spinning around on the surface of the water in bewildering and rapid gyrations are whirligig beetles. Oval, shiny, blue-black to dark brown, the beetles have a divided eye that is believed to enable them to see both above and below the water. The middle legs, and more especially the hind legs, are adapted for swimming, being greatly flattened and fringed.

Come autumn, most of these beetles will fly at night in search of new habitats.

July, the crown of summer, slowly falls to listless days, with the green of May becoming tarnished and June's choral singing rendered silent. It is the month when young birds are aplenty, discovering the wonders of flight. Adults begin to molt. Nests empty and will soon fall apart. Butterflies, dragonflies, and damselflies take wing to maximize their short lives. Seeds mature in all shapes and sizes. The year is now lusty and full grown, looking ahead as it prepares for the next spring.

A late summer view of Upper Glebe Marsh, from the Observation Deck.
Jug Bay Wetlands Sanctuary

AUGUST
The Thunder Moon

All your renown is like the summer flower,

that blooms and dies; because the sunny glow,

Which brings it forth, soon slays with parching power.

Dante Alighieri, *Purgatorio*, canto XI, 115–117

1 August | PHRAGMITES: TALLEST OF THE TALL AND THE ONLY TRUE REED

Towering stands of the feathery, plumed seedheads of phragmites sway in the wind amid portions of the marshes bordering the Patuxent River. This grass, with a relatively minor presence in the river, has become a major pest elsewhere. Indeed, it may be one of the most widely distributed seed plants in the world.

Since the 1930s, phragmites has slowly increased its range in Chesapeake Bay and its tributaries, choking natural vegetation, changing the hydrology and, thereby, wildlife habitat. Its thick, eight-foot stems block light from reaching other plants, and its rootstocks (rhizomes) outcompete native plants such as cattails, rapidly turning large areas into a phragmites monoculture.[1] The resultant growth eliminates small intertidal channels and obliterates pool habitats that offer refuge and feeding grounds for invertebrates, fish, and waterbirds. It also creates a dense jungle of vegetation that native birds, furbearing mammals, and even white-tailed deer cannot penetrate. In addition, decomposing phragmites can raise the surface elevation of the marsh more rapidly than the slower-growing native marsh vegetation—not a bad development in some spots, given rising sea levels.

Genetic studies, such as by Kristin Saltonstall and colleagues, have confirmed that there is a native subspecies of phragmites found in some areas of Maryland and other parts of the United States. In the early nineteenth century, nonnative phragmites, most likely of European origin, appeared in coastal ports in the eastern

United States. Its marked expansion may be due to its more vigorous growth, aided, in the twentieth century, by the construction of railroads and major roadways, habitat disturbances, shoreline development, pollution, and eutrophication.

Once established, phragmites becomes almost impossible to eradicate. Although many prefer to avoid the use of chemicals, herbicides provide the most effective primary method of control, particularly when coupled with nonchemical methods that further stress this invasive plant.

Birds, such as common yellowthroats, marsh wrens, swamp sparrows, and least bitterns roost in phragmites, and red-winged blackbirds, as well as some wading birds, have been documented to nest in stands of this grass. Muskrats are the only mammal species to feed on phragmites to a noticeable extent.

2 August | Full-Throated Ease Gives Way

Further in Summer than the Birds

Pathetic from the Grass

A minor Nation celebrates

Its unobtrusive Mass.

Emily Dickinson, "Further in Summer than the Birds"

As ever, a poet claims our thoughts. With birdsong no longer filling the golden hours of summer, most species are now tending their offspring, and some are preparing to migrate. In its place, the air is saturated day and night with the calls of grasshoppers, crickets, cicadas, and katydids.

Through their chirrups, peeps, trills, buzzes, shuffles, crackles, and clicks, the males hope to attract females. Crickets and katydids rub their wings together (called stridulation), using an uneven ridge, or file, on one broadened front wing and a sharp edge, or scraper, on the other to produce a tone like that of a fiddle. At the same time, the thin, papery portions of their wings vibrate, amplifying the sound. They hear their songs through a pair of ears (tympana) on each foreleg, or tibia, just below what we would call the knee of the insect.

Not surprisingly, crickets produce different songs for different reasons. The calling song, heard for distances of up to a mile, aids a female in finding a male. Once she is nearby, the male switches to a courtship song to convince her to mate.

In some instances, the male sings a postcopulation celebratory song. Crickets also chirp to establish their territories and defend them against competing males.

Grasshoppers also stridulate. In flight, males (and sometimes females) make loud snapping or crackling sounds with their wings, especially during courtship. This unique sound is called crepitation, being produced when the membranes between veins are popped taut.

In contrast to crickets, katydids, and grasshoppers, the sound of the cicada's courtship song can be deafening. Males possess two ribbed membranes, called tymbals, on each side of their first abdominal segment. Contracting their tymbal muscles causes their ribs to bend suddenly and produce sounds that are amplified in abdominal air sacs. For many species, songs start slowly and build to a crescendo before dropping off. Males often aggregate as they sing, creating the familiar cicada chorus.

A female cicada attracted to a mate responds by a maneuver known as a wing flick. The male, in turn, replies with more clicking of his tymbals. As the duet continues, the male makes his way toward her and begins a courtship call.

My initial search for these denizens begins amid bushy areas at the wetland's edge, where I detect the loud rattling trills of some trigs (crickets), possibly two-toned red-and-black handsome trigs. Then, after a prolonged search in the meadow, I run down the beautiful owner of a sonic burst of ticks, followed by a loud, buzzy trill—a handsome meadow katydid.

In the fading light I wander to some rocks near the Wetlands Center, where field crickets offer unmistakable chirps, and then to the nearby meadow. As both crickets and katydids use vegetation to bounce their sounds, it is by accident that I stumble on a Carolina ground cricket, a familiar companion to the grigologist (someone who studies crickets, katydids, and cicadas). Its song, for a small creature, is surprisingly loud. It is likely to be the last ground cricket singing in the face of the coming winter.

Much as I search (now by flashlight), I cannot uncover the wonderfully named tinkling ground crickets and confused ground crickets (their common name results from observers initially confusing them with Carolina ground crickets). But I do hear mole crickets singing from their burrows, carefully constructed with horn-shaped entrances. These help amplify their songs, projecting them skyward, to be

heard at some distance. After mating, the female ejects the male from the burrow
and takes possession of it to lay her eggs.

At dusk a great blue heron flies across the rising moon and fading pink sky.
The chorus of crickets and katydids has no competitor and will continue into the
night.

3 *August* | BIRD WALK, FLOWERS, AND MUSHROOMING

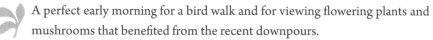 A perfect early morning for a bird walk and for viewing flowering plants and
mushrooms that benefited from the recent downpours.

A delightful patch of bouncingbet (or soapwort), bearing clusters of little
white flowers, abuts the railroad trail. As the latter common name implies, when
boiled in water, the leaves yield a very gentle soap. Colonists used this nonnative
plant to clean delicate fabrics or textiles, and it has been hypothesized that it was
used to treat the Shroud of Turin. The name "bouncingbet" is thought to come
from the visual aspect of the downturned petals, looking like the bent-over rear
of a washerwoman named Bet. The plant is also called goodbye summer.

Not far away, in wetter conditions, are two-foot-high fringed loosestrifes, with
starry yellow flowers nodding on long slender stalks. This common name also
bears an interesting history, referring to the plant's supposed power to sooth animals, or loose (free) them from their strife. Legend has it that as Lysimachus, the
King of Sicily, was walking through a field, a bull started to chase him. He grabbed
a loosestrife and waved it in front of the bull, eventually calming it.

A delightful sight is the inconspicuous but tenacious nakedflower ticktrefoil,
with tiny pink to bright purple flowers and bright green seeds. The flowers turn
light blue after pollination.

The marsh vegetation has reached it full height, and seven-foot-tall colonies
of crimsoneyed rosemallows wave in the gentle breeze, showing their pure white
petals with a deep maroon center. Many are past flowering and bear golf ball–sized,
beaked seed capsules.

In the midlevel marsh stand large, bright green areas of 12-foot-high wild rice.
Female flowers top the broom-like branches, while male flowers dangle from
spreading branches below. Their thin, half-inch-long black seeds have long tails
that help them drop directly into the mud when ripe, where they germinate the
following year.[2]

Unlike the wild rice of some northern states, the seeds of wild rice in the Mid-Atlantic region have no commercial value. Game birds, songbirds, muskrats, and waterfowl eat them, however, and wood ducks consume both the seeds and the flower petals, as do red-winged blackbirds, now flocking to the marsh in impressive numbers. Marsh wrens attach their woven, globular nests to its tall stems and, as we will witness in the autumn, wild rice areas are a stopover point for soras.

Cattails also form extensive stands, easily recognized from now until winter by a brown, cigar-shaped "cat's tail" made up of thousands of tiny seeds. The seeds have their own long tails, allowing them to be dispersed by the wind. Cattails provide shelter for waterfowl, red-winged blackbirds, soras, and marsh wrens, and muskrats use them to construct dome-shaped lodges in the marsh. Muskrats also often clear a path through a cattail stand, a track sometimes used by other wildlife. Paradoxically, cattails trap sediments in the water column and slowly change their habitat from wet to dry, creating conditions eventually proving unsuitable for their survival.

Following the heavy rains of the last few days, mushrooms are a'popping. I know little about mushroom taxonomy and, since there is no descriptive list for Jug Bay, I have to use available field guides. Black-footed polypores, growing in a group on decaying deciduous wood, are easily identifiable. They bear reddish-brown caps and a black stalk. Nearby, on a fallen oak, the tops of beefsteak polypores look like slices of prime beef (but are disappointing to taste), although these mushrooms are creamy white below.

Farther along, I find honey mushrooms at the base of a deciduous tree. They sometimes behave like a parasite, for once their black strands (rhizomorphs) penetrate the bark of trees and shrubs, they produce a branching network of white, lace-like mycellium (filaments) that spread up the trunk or branch, eventually leading to the death of the host plant. Their yellow-brown caps can grow up to four inches across, and the stems, up to six inches long.[3] These mushrooms are edible, as long as they are well cooked, but there are poisonous mushrooms that have a very similar appearance.

I enter a grove of maples and find a strikingly beautiful netted rhodotus mushroom, with a two-inch, reddish-pink, convex cap netted with white ridges and pits. In a wetter area, on dead hardwood, is a group of fungi reminiscent of coral reefs—a cluster of largish crown-tipped coral mushrooms, multiple-branched yellow structures with cup-shaped, crown-like tips.

A coral mushroom. *Kerry Wixted*

Such has been the frequency of cloudy skies and rain that fairy rings of mushrooms now grow in woods, parks, and lawns. While the rings are fleeting, their underground fungal mycelium grow for many years, periodically popping up mushrooms. As an initial spore sends out its mycelia in all directions, a ring slowly forms.

I squat to take a closer look at some mushrooms but am distracted by a movement in the decaying leaves. I find a well-hidden Fowler's toad, red rust in color, with darkened wavy spots. As well as using camouflage, these toads also secrete a noxious compound from the warts on their backs. They are also known to play dead.

Fowler's toads reproduce repeatedly in warmer seasons of the year, especially in May and June, breeding in the shallow marsh. The male's call attracts females and other males, the latter sometimes attempting to mate with the male making the sound. The vocalizing male then produces a chirping release call, informing the other male of his mistake. When the vocalizing male eventually finds a female, the pair embrace, and 7,000 to 10,000 eggs are fertilized. The eggs hatch into tadpoles in two to seven days. In one small pond, as many as 10 different age classes (separated by several days) were observed over the course of a breeding season.

A Fowler's toad. *Rob McEachern*

A new tadpole may reach sexual maturity in one season, but this has been known to take up to three years.

I leave the Sanctuary with a pair of juvenile Cooper's hawks calling and thrashing around in trees and shrubs, behaving as adolescents are wont to do.

4 and 5–6 August | Cool Air and Summer with a Difference

Several masses of cold air dropped into the Upper Midwest on 4 August and brought in a period of below-normal temperatures for the next 10–14 days for the eastern two-thirds of the country. Some of the cold air, more typical of late September or early October, spread toward the Mid-Atlantic region and the Northeast during 5 and 6 August, providing a perfect summer weekend, with drifting puffy clouds garlanded by clear blue skies. Maximum temperatures reached 81°F (some seven degrees below the norm) and are expected to be followed by five successive days of 90°F or more, some of them turning stormy.

This pattern is expected to persist through mid-August and reverse later in the month, bringing in warmer than normal temperatures in the eastern two-thirds of the nation.

5 August: Four Square Kilometers of Planet Earth

I have always had a soft spot for amphipods and was delighted to learn that a rare species has been found again in Rock Creek Park in Washington, DC—its only home in the entire world. The milky white, blind, one-centimeter-long Hay's Spring amphipod was rediscovered in wet dead leaves, having been detected by environmental DNA, a technique for examining life in soil and water to uncover tiny creatures. Scientists took samples from seepage springs and discovered these amphipods in places where they were not found by traditional methods.

The existence of Hay's Spring amphipods helped promote a program to control stormwater runoff and has been cited as a reason to stop the construction of the light rail Purple Line along the northern boundary of Washington, DC.

Source: Nemiller et al. 2017.

Much of the summer is past, and on 5 August, the 90 longest days of the year came to an end. Autumn beckons.

News flash: The beaver dam at Beaver Pond burst, due to heavy rain on the weekend. A similar breech in 2011 took two months to repair.

10 *August* | The Names They Have

 New flowers are emerging and adding pleasure to a summer's day. I find further delight in their often-strange names.

A Carolina elephantsfoot's flower is truly spectacular. It is native to moist, woody areas and derives its common name from the large flat leaves, resembling the foot of an elephant. Such leaves can be as large as 8 inches long and 3.5 inches wide. The flower does not have true petals, an equally unusual characteristic. Instead, there are several disk-shaped small flowers of pale violet. In late October, the lobes of each of these florets will dry and turn into brownish-maroon seed clusters.

The common name of another plant, wingstem, comes from its distinctive stem, having vertical ridges, which are sometimes described as wings. Its flowers grow in loose clusters at the top of the plant and look like small, bright yellow sunflowers, with the petals drooping haphazardly. They are visited for their nectar primarily by "friendly" long-tongued bees, especially bumblebees, while several kinds of insects feed destructively on its foliage. Caterpillars of silvery checkerspot butterflies eat the leaves; those of gold moths feed on the flowers and developing seeds; and the caterpillars of two gracillariid moths (taxonomic family Gracillariidae) perform leaf mining. Other insects consuming wingstems include leaf beetles, the larvae of some gallflies, and certain aphids, including green stink bugs. Clearly, the dense colonies of wingstems offer a bonanza for insects, though the bitterness of its leaves makes this plant unpalatable to white-tailed deer, rabbits, and other herbivores.

Common bonesets, or thoroughworts, are tall, erect plants with flat-topped clusters of small composite flowers and opposite, or whorled, toothed leaves. The common name for this species may come from its use to treat what was known as breakbone fever (dengue fever). In addition, it is often used as a diuretic, and the leaves of some boneset species were once considered to be useful in mending broken bones, though this may originate from a theory known as the doctrine of signatures.[4]

The name "thoroughwort" refers to how the stem appears to pierce the leaf.[5] Seeds of this plant come equipped with a hairy papus—a parachute-like structure—that is an efficient means of distributing the seeds by wind. Butterflies, bees, and flies are the plant's foremost pollinators, and it is renowned for attracting ambush bugs, which lie in wait for their prey on flowers or other plant parts.

A neighbor of the bonesets, inhabiting a partially shaded marsh wedge, is sweetscented joe pye weed (trumpet weed), with large, rough leaves borne on upright thick, round, purplish stems. From midsummer to early autumn, the nectar in its clusters of white-pink or pink-purple flowers attracts butterflies and bees. White-tailed deer browse on the leaves, while swamp sparrows eat the seeds.

The common name refers to Joe Pye (*Jopi* in the Native American tongue), an Indian healer from New England who used this species to treat a variety of ailments. Folklore states that American colonists used this plant to treat typhus outbreaks.

12 August | HAWKMOTHS

Moths get little respect, except from the relatively few scientists and naturalists passionate about their dramatic behavior and features. Yet an estimated 11,000

moths are known to occur in the United States, and another 160,000 species have been catalogued from around the world. Few people know that the number of moth species is far greater than the world's species of butterflies (17,500). Fewer still are aware that far from being a drab brown or white, many moths come clothed in myriad colors and patterns, some brighter than those flashy butterflies.

Hawkmoths are a group that invariably draws attention, in the same way that orchids engage nonbotanists. These magnificent creatures have long, narrow fore-wings and bullet-shaped bodies. They are fast flyers—erratic, yet often highly aerobatic—and they can hover. Some can briefly fly backward or dart away, much like hummingbirds. Many are experts at finding sweet-smelling flowers after dark.

The species found in Maryland include the beautiful pink-spotted hawkmoths, the green-and-burgundy-backed hummingbird clearwings, and blinded sphinxes, with bright pink hindwings. All have an extraordinarily curved tongue (proboscis), as long as the entire insect, that can easily reach down to a flower's deep nectary (the gland that secretes nectar).

In his book on the fertilization of orchids, first published in 1862, Charles Darwin examined orchids that had nectar spurs (projections of various lengths that allow pollinators to land on the elongated tissue and more easily reach the nectaries) nearly a foot long, but he was ridiculed by other scientists of his day for predicting that these orchids would be pollinated by moths. After his death, hawk-moths with tongues long enough to sip the nectar produced by Star of Bethlehem orchids were discovered on the island of Madagascar and elsewhere.

21 *August* | THE SOLAR ECLIPSE AND WILDLIFE

My son and I are fortunate to observe the pulsing and dancing ring of fire of a total solar eclipse this day in Oregon (for Ian) and Colorado (for myself), and I ponder the impacts on wildlife. It is a rare event, with a coast-to-coast swath of totality. Reports suggest that a total eclipse appears to have no profound long-term effects on wildlife, producing only temporary alternate behaviors: birds stop singing and return to their roosts, locusts and cicadas halt their calls, spiders begin undoing their webs, and squirrels retreat to their drays. Experts say that most wildlife resume their normal activities once a total eclipse begins to wane.

Data from six red-tailed hawks, fitted with GPS equipment, may help research workers from the University of Nebraska determine the effects of a total solar eclipse. Some information had been collected a week ahead of Monday's eclipse and more is expected to be gathered during the week following the eclipse. A question the investigators hope to answer is whether some animals can sense the approaching alignment of the Earth, moon, and sun hours before the eclipse starts. Other questions include whether nocturnal hunters, such as owls, wake up in the middle of the day, or if other birds go to their roosts when conditions simulate dusk and return to their daytime routines after the eclipse.

Purdue University's Center for Global Soundscapes will record how the eclipse will affect animal habits and vocalizations and ask various questions, such as are some of the crickets that sing at night going to start singing during the middle of the day, and is the northern cardinal that is normally singing during the day going to stop singing? The noises animals make—or don't make—will offer a chance to gather clues on the role of light in normal animal activity and will allow researchers to study how unexpected darkness may alter an animal's circadian rhythm—its 24-hour cycle. As one researcher notes "Light availability influences the daily behavioral patterns of animals. We anticipate a general confusion among animals, including birds, frogs and bats. We just don't really know because it is not well documented on this large of a scale."

The investigators are hoping to collect thousands of audio files from the eclipse's line of totality, as well as from areas where the sun is only 60 percent eclipsed. The audio files will be recorded at all stages of the eclipse—before, during, at its peak, and after—and indicate where the recordings were made.

The California Academy of Sciences has created an iNaturalist app and invited citizen scientists to take pictures of eclipse-related animal behavior. The value of citizen science is demonstrated yet again.

23 August | Summer Heat

The high temperatures these last few months bring benefits and costs to natural habitats and their wildlife. Warmer conditions are normally critical for the survival of plants and animals, but the ensuing competition can be intense. Trees and other plants struggle to reach sunlight and crowd out smaller competitors, often imperiling the latter's survival. Animals are also affected, particularly small organisms. Under moist conditions, summer heat stimulates the growth of bacteria and viruses, spreading disease. These conditions can also enhance the viability of insect eggs and thus increase adult insect populations, which smaller animals eat, thereby passing more energy through the food chain.

Birds—with a high body temperature (105°F), a high metabolic rate, effective insulation, and limited means to get rid of excess warmth—are vulnerable to overheating in very hot weather. They can open their bills and flutter their throats to allow moisture to evaporate from their mouths, reduce insulation by compressing their feathers tightly against their bodies, hold their wings away from the body to expose bits of bare skin, swell bare patches to increase the surface area of their bodies, and increase blood flow through their bills to dissipate heat. Other strategies birds use include behavioral adaptions, such as panting, adopting suitable activities, seeking shade, soaring to higher altitudes, bathing, spreading their feathers to circulate air across hot skin, turning their lightest-colored plumage toward the sun, and retreating to cooler locations. In a distributional response, some birds are moving their breeding range to more-northern latitudes, to keep pace with climate change.

Parent birds stay active and remain out in the open during the heat of the day, since, with young in the nest, they must continue feeding their growing offspring. In desert areas, some birds have evolved remarkable adaptations. For example, common poorwills (small nightjars) can dissipate up to five times their metabolic heat production. They can also become torpid when incubating their eggs.

In the Jug Bay area to date, there have been 40 days of 90°F temperatures against an average of 31 days and a yearly average of 36 days. In 1980 and 2010, 67 days of 90°F temperatures were recorded.

The amount of rain that has fallen thus far stands at 30 inches, against an average to date of 27 inches.

25 August | RAINFALL

Rain has its own language: sometimes aggressive, sometimes gentle. Either way, this year it has shortened the leafiness of summer. Battered leaves are now falling, and vegetation is assuming a beaten aspect. Hay is being gathered, and mists are now rolling off the Patuxent River and its sinuous creeks.

Many people find the scent during and immediately after a rain to be pleasant or distinctive. One source is petrichor, a plant oil absorbed by clay-based rocks and soil. In the 1960s, Australian researchers I. J. Bear and R. G. Thomas described how, during rainfalls, the oil is released into the air. Another source of the distinctive scent is geosmin (an organic compound produced by certain bacteria), which has a warm, earthy smell.

More recently, Macrina Cooper-White has described how scientists at the Massachusetts Institute of Technology used high-speed cameras to record how the scent moves into the air. When a raindrop lands on a porous surface, air from the pores forms small bubbles, which float to the surface and release aerosols. Such aerosols carry the scent, as well as bacteria and viruses from the soil. Raindrops falling at a slower rate tend to produce more aerosols, explaining why petrichor is more common after light rains.

Some scientists believe that humans are sensitive to this scent, as our ancestors may have relied on rainy weather for survival.

Over the contiguous United States, total annual precipitation has increased at an average rate of 6.1 percent since 1900, with the greatest upswings occurring within the east-northcentral climate region (11.6%) and the South (11.1%). Hawaii was the only region to show a decrease (9.25%).

Climate Central's analysis of 65 years of US rainfall records shows that the lower 48 states have experienced more incidents of heavy downpours since 1950. The largest increases are in the Northeast and Midwest, which, in the past decade, have seen 31 and 16 percent more heavy downpours, compared with the 1950s.

Former names for a heavy or violent rain include gully washer, trash mover, toad strangler, cob floater, duck or goose drownder (drowner), lightwood knot floater, sod soaker, and sheep drowner.

26 *August* | LANDSCAPE AND THOREAU

Tidal flows, passing clouds, and restless trees constantly recreate themselves, even as we struggle to capture them. This is one reason why we return to a landscape with such unalloyed pleasure. We feel it to be permanent yet changing, ageless yet time bound. Poets struggle mightily to capture what they see, and painters can only emulate passing moods, at best. Nonetheless, schools of artists made landscape painting reach its golden age in the nineteenth century, and it is to them that we turn for assurance, as well as a deepening of our appreciation as we take time to observe both relationships and the whole.

If we ease gently into a landscape, we can admire its form and features, as well as its geological underpinnings. We can observe the sky and the weather shaping the land, the plants and animals that populate its many habitats, and man's imprint on it all. With the elements pieced together, the ordinary becomes extraordinary, and we are often enamored. But change is ever the case, for there is no definite beginning or end. In capturing a moment, we take comfort in knowing that we are fully sentient and that our communication with nature speaks to an inner voice.

I muse on this matter and am surprised that Thoreau's 200th birthday on 6 July drew modest attention in the media, save for reviews of old and recent biographies and an article on the famous Walden Pond. His writings are part of school curricula, and he helped lay the groundwork for what has become the discipline of ecology. He was one of the first advocates for the establishment of a system of national parks, and his passionate championship of the ethical treatment of all living things heralded environmental conscientiousness. Asked once why he was so eternally curious about such things, Thoreau responded, "What else is there in life?"

He was devoted to science, chronicling the dates when plants flowered (an important record as the climate changes) and performing groundbreaking research into the succession of tree species in burned and logged forests. He linked science, politics, and an appreciation of nature and fostered the hope that mankind and the natural world might flourish together. It was a unique view then, although it continues to this day. Thoreau, in his essay "Autumnal Tints," instructs us thus: "Objects are concealed from our view not so much because they are out of our visual ray as because there is no intention of the mind and eye toward them.... There is just as much beauty visible to us in the landscape as we are prepared to appreciate, not a grain more."

28 August | Underwater Plants, Sedges, Rushes

Underwater grasses, more accurately termed submerged aquatic vegetation, or SAVs, are much in evidence at this time of year. Growing entirely underwater and relying on buoyancy to support their stems and leaves, some also have flowers or tufts that poke slightly above the water's surface. These plants help remove harmful nutrients and sediment pollution, stabilize sediments, reduce wave energy and erosion, and enrich shallow aquatic environments. They also provide refuges for fish and sustenance for waterfowl.

Coon's tail (or hornwort) and waterthyme (or hydrilla) are common examples in the Jug Bay area. Both are invasives, having been introduced as aquarium plants. The former species, with its bushy leaf clusters (like a raccoon's tail), has allelopathic qualities: it excretes substances inhibiting the growth of phytoplankton and cyanobacteria (blue-green algae). Significant populations of worms, snails, crustaceans, and insect larvae occur in these clusters, however, and are much sought after by other wildlife. Waterthyme also produces compounds that inhibit the growth of native species, and its dense growth can outcompete native underwater vegetation, leading to a loss of biodiversity. Nevertheless, both are an excellent food source for waterfowl and a habitat for fish.

Another invasive species is curly pondweed, noted for its numerous pencil-like flower spikes, produced from a branching stem of flattened leaves up to three feet long. It has a double arrangement of leaves: the submerged ones are grass-like undulating streamers moving in concert with the water flow, whereas its floating leaves are rigid lily pads.

SAVs are one of the most important plant communities in the Patuxent River and the Chesapeake watershed. In the last few decades, however, Chesapeake Bay has suffered a dramatic decline in SAVs, both in their overall acreage and in species diversity, due to pollution, siltation, and excessive amounts of nutrients in the water.

Sedges and rushes have always held my grudging attention. Though they are very difficult to identify, their solid stems and fruit capsules are intriguing. The following rhyme applies to plants occupying the edge of the marsh:

Sedges have edges,
Rushes are round,
Grasses are hollow,
What have YOU found?

Sedges constitute one of the largest genera (*Carex*) of vascular plants and are the very devil to classify, with some 29 species inhabiting Jug Bay and over 400 in North America. They occupy many habitats, ranging from tidal and nontidal wetlands to meadowland, and the different species are best told apart by their flowering or fruiting spikes, giving rise to somewhat telltale names—bottlebrush, smoothsheath, drooping, awlfruit, prickly, and blunt broom. Most sedges grow in wetlands, but there are many in woodlands and grasslands. The roots and rhizomes of sedges harbor more biomass than the stems and leaves. This may explain why they are pioneers in forming shoreline mats that, in time, seal off water entry and form a succession of wetland types: marsh, shrub swamp, forested swamp, and upland.

Rushes are more easily recognized by their smooth, round, spire-like stems, grass-like leaves, and three-part flowers and seed capsules. Nonetheless, like sedges, identifying rush species requires an examination of their highly complex flowers or seed clusters. Of the six species at Jug Bay, the common, or soft, rushes are the most prevalent, occupying the water's edge in large clumps some four to five feet tall. Rushes in general often form adjacent colonies with softstem bulrushes and frequently cohabit with curly pondweeds, common duckweeds, and pickerelweeds. More particularly, they all are hosts for many insects and often form spawning grounds for fish. On occasion, one may see nymphal dragonflies ascending a stem to transform into their winged form.

31 August | BIRDING MONTH

For many, August ranks among the richest of birding months. Southbound shorebirds are near their peak diversity, and many songbird species have begun migrating between North America and tropical climes in the south, their flight buoyed as the winds turn northerly.

So it is with some expectation that the twice-monthly bird count gang sets off from the parking lot of the Sanctuary. From the Observation Deck, we can see the now browning cattails and the drooping, mud-flecked spatterdocks and contrast them with the vivid green of the 10-foot-tall extensive beds of wild rice, bearing seed-producing flowers. Clouds of red-winged blackbirds are feeding on the immature "milk" seeds of wild rice and seeking the cover it affords during their molt—missing flight and tail feathers are a great handicap. Some birds move up the stalk until it bends over to reach a seedhead, while others move sideways, bending the stalk horizontally as they approach the seedhead. Sometimes the lighter-weight females hover at the seedhead and grasp a seed.

Mist rolls off the Patuxent River, revealing and then hiding lesser yellowlegs and a pair of wood ducks. The alabaster white of southbound Forster's terns and a sole great egret draw us like a magnet, and we are only pulled away from this specta-

Prothonotary Warbler Breeding Success

To many a birdwatcher, there is nothing more pleasing than hiking through a forest and hearing the loud "tsweet-tsweet-tsweet-tsweet" of the prothonotary warbler. Originally named for the prothonotaries, or clerics, in the Catholic Church, who wore brightly colored robes, the bird is often called the golden swamp warbler. It breeds in the eastern United States and overwinters in Mexico, Central America, and northern South America. It is one of only two North American warblers to use a cavity nest site.

The Monitoring Avian Productivity and Survivorship (MAPS) bird-banding program at Jug Bay has captured and banded 51 prothonotary warblers since the program's inception in 1990, an average of 2.8 new birds captured per year, according to Mike Quinlan, a long-time bander with the program. In 9 of the 18 years, however, no prothonotary warblers were banded, and the average number that was banded has decreased significantly: to 1.3 per year in the 10 years from 2007 to 2016. Sadly, North American Breeding Bird Survey data show a loss of more than 50 percent of the population over the past 50 years, primarily due to habitat loss.

In spring 2014, Frode Jacobsen introduced plastic pipe nest boxes at Jug Bay to establish a population of these warblers. As nesting in them proved to be unsuccessful, the boxes were replaced with wooden ones in 2015/2016. As this again saw no nesting success, in March 2017, Alan Christian fitted nine of the original boxes with eight-inch stovepipe predator guards and installed them over tidal marsh on the Otter Point peninsula, along the Otter Point and Two Run Trails. Nest building began in the beginning of May, with the first eggs being observed by 23 May. The last chicks fledged by 3 August.

Alan observed six nests with a total of 28 eggs (4.6 per nest), and 27 young fledged (4.6 per nest). With the simultaneous nesting activity in June and July, it appears that at least four pairs of prothonotary warblers had bred successfully, utilizing the boxes. The two pairs that began their breeding activity in May were considered to have a second successful clutch; the two pairs that didn't start nesting until mid-June only produced one brood for the season.

While it could be a coincidence, the MAPS exercise banded five prothonotary warblers in 2017.

cle by the majesty of a juvenile bald eagle sweeping across the landscape. But such is the life of birders, especially when another competitor grabs our attention—a northern harrier, its long wings allowing it to hug the marsh as it searches for prey. It is the first to appear in the fall season. To the surprise of all, a double-crested cormorant glides into view, doubles back, and lands midchannel. Within seconds it has secured a fish.

While most ducks are late fall migrants, some species are on the move in August. Therefore, we are not too surprised to learn that a blue-winged teal was recorded earlier in the week.

We walk the woodland trail, aware that the crunching of leaves alerts all wildlife. Vigilant eyes spot a northern parula warbler, an eastern wood-pewee, an eastern phoebe, and a singing red-eyed vireo, all doubtless dispersing as much as migrating. The scarlet leaves of blackgums break up the green foliage, and ruby-throated hummingbirds visit cardinalflowers on the marsh's edge.

At the side of the Railroad Bed Trail, many tiny, pinstriped caterpillars feed ravenously on a greenbrier, their only host plant. When someone touches the plant, the caterpillars vibrate in unison. They are turbulent phosphila, casually known as "two-headed caterpillars," with the false head they present to predators being assumed to be a defense strategy. Some lepidopterists wonder if this insect species is chemically protected or if its patterning is largely a ruse.

On the walk to the Observation Blind overlooking Lower Glebe Marsh, I notice that the cattails are well-advanced in sporting brown, sausage-shaped seedheads and beginning to shed seeds that are attached to downy parachutes.

North American beavers are busy forming a mound of vegetation before the Observation Blind. At some three feet above the waterline and six to seven feet across the base, it is truly impressive. Though muskrats are most active at night or near dawn and dusk, we see the tail of one whipping back and forth like a snake and, to our amazement, being used to steer backward. Because of its flattened tail, the Old English name "musquash" became altered to "musk-beaver"; later it became "muskrat," due to the animal's resemblance to a rat.

Migrating Canada geese are back (briefly joining their resident relatives), and we enjoy their comforting calls as they fly southward. During the hunting season, those choosing a rest stop seek the safety of the Sanctuary. Ospreys remain, but will they be here for long? Happily, barn swallows and tree swallows continue to flit over the marsh.

The Butterfly Garden at neighboring Glendening Preserve was jumping today, with 16 species, including several monarch, sleepy orange, and cloudless sulphur butterflies; three different kinds of swallowtails; and skippers massing on zinnias and lantanas. Some 10 years ago, Dave Perry led a team of volunteers in establishing the garden, now attracting many such species, as well as butterfly enthusiasts from across the region. More recently, Darcy Herman has conducted a study on the flight and crawl seasons of butterfly species as part of the USA National Phenology Network's citizen science program, monitoring the impacts of climate change.

I leave the butterflies and their dazzling displays and admire the balloon-like movements of dandelion and milkweed seeds eddying on the slightest breeze.

August announces that the year is retreating, marching ahead to the slow but steadily looming winter. The time of growth is past, and nature is now bent on ensuring plants and animals will be conveyed safely through harsh times, able to greet the spring. Breeding is largely over. Some summer bird visitors are provisioning for their long southward flights, while residents continue molting and begin their dispersal into their respective habitats.

The edge of Lower Glebe Marsh in the fall, showing the seasonal dieback of vegetation. *Rob McEachern*

SEPTEMBER

The Hunting Moon

Sorrow and scarlet leaf,
Sad thoughts and sunny weather,
Ah me! this glory and this grief
Agree not well together.

Thomas William Parsons, "A Song for September"

1 *September* | CLOUDS

How often throughout the seasons do we look at clouds, admiring their form and the way they capture dawn's early light, build to a storm, or hold the dying embers of sunset? They keep us in rapture or fear, never losing their mystery or majesty. Art, literature, and science provide their own visions of these aerial landscapes, often rooted deep in our imagination, such as in Wordsworth's poem that begins "I wandered lonely as a cloud / That floats on high o'er vales and hills."

It was Luke Howard, in 1803, who first named the types of clouds, a classification that remains with us to this day: cirrus, stratus, cumulus, and nimbus. Indeed, we may take comfort in these regular features in otherwise disconcerting times. He once wrote: "The ocean of air in which we live and move, with its continents and islands of cloud, can never to the conscious mind be an object of unfeeling contemplation."

So as we gaze upon the sky and its weightless forms of air, we must thank Howard for deepening our relationship between the earth and the ever mutable, overarching sky. He would doubtless have been intrigued to learn that clouds play a major role in regulating climate and climate change. They screen sunlight from the Earth and are critical in accelerating the radiation of heat away from it.

In a *Washington Post* article by Amy Ellis Nutt, Gavin Pretor-Pinney, founder of the Cloud Appreciation Society, says that we should not forget the beauty and impact clouds have on our lives, since "cloudspotting is a conscious invitation to daydream, a sensitivity to your surroundings."

3 *September* | FURTHER CHANGE

Northerly winds and cooling temperatures make September a month of change. Though 21 September is the official start of autumn, the natural world is already responding. Ragged and pockmarked leaves are being shed, assuming yellow, orange, purple, and fiery red. Seeds are well developed (even American holly berries are turning red). Highways and byways sport the gold of great ragweeds and wreath goldenrods, and the ripe downy gray knobs of common dandelions and hawkweeds are wafted far and wide by the wind. In the meadow, the thistle-like, dotty purpling of spotted knapweed flowers flourish atop small pineapple-like structures, wiry stems, and sketchy leaves.[1] A medley of colors vie in the showery wind.

The shower passes, and in the sunshine, bees, hoverflies, and butterflies mass and resume their search for knapweed nectar. The air is again filled with the monotonous sound of insect music. In the neighboring woodland, the golden brilliance of sunlight forms pools of incandescence.

The stalks of wild rice have bronzed with age and the seeds will shortly turn purple black.[2] Beside the marsh, the long pods of purpleleaf willowherbs have coiled back like springs, and feathery slender hairs or bristles replace purple blossoms, in order to carry the seeds far and wide. Concurrently, new shoots spring from the bases of the stems, heralding renewed growth from the rootstock to sprout next spring. Colonies of waterpeppers stand tall, their fruit brown as coffee. In dying, all these plants will become brown and withered.

Monarch butterflies are moving south, and birds in the region push through the Atlantic Flyway. Some waterfowl will remain in Chesapeake Bay and water bodies in the bay's watershed, where they will become the targets of hunters or of scoping by birders. Songbirds now sport duller feathers, which, in the case of warblers, present a great challenge to their identification. Mammals begin to gorge on nuts and insects. Amphibians are moving to winter homes, and fish born in freshwater, such as gizzard shad and alewives, begin their migration to the Atlantic Ocean. Katydids and crickets call in the lengthening darkness, their frequency and strength reflecting the declining temperatures.

Sunlight is softer, no longer the summer's harsh glare. Sunbeams gently illumi-

Wild rice, showing the grains beginning to mature. *M-NCPPC Department of Parks and Recreation*

nate the gossamer strands of a spider's web. A capsule of down twists in the web, without hope of escape.

In the cool breeze, the dog days of summer are ending.[3] But with a plenitude of spiders and bees, and midges newly emergent from the marsh, passing swallows and warblers feed on them as they wend their way south. The wondrous landscape is visibly changing before the advance of autumn.

5 *September* | SPOTTED TURTLES

The sun has warmed the day. The crickets' song is loud, some birds call, and I see several large painted turtles basking on a log in the pond behind the beaver dam. Insects inspect the purple flowers of pickerelweeds and the red-pink of waterpeppers, or swamp smartweeds. A closer step to the pond's edge allows me to glimpse the sudden movements of a hatchling spotted turtle as it rapidly disappears beneath the plushy weeds.

I have patience, and in time it reemerges, its inch-long upper shell garlanded by a string of algae. The turtle has journeyed from a nest its mother created some three months ago. Its carapace is black, with one yellow spot on each segment. Eventually, more spots will appear and it will truly deserve its name. If I was to turn it on its back, a deep golden yellow would be revealed, edged by brilliant red.

Its mother probably chose an open nesting site, such as at the swampy edge of the forest. After digging a small nest with her hind legs and feet, she laid two to seven eggs and, in contrast to most turtles, covered them to disguise the nest. In three to four months, the hatchlings would have emerged and headed for wet, grassy areas in search of food and shelter, eventually reaching shallow aquatic habitats. On their terrestrial journey, these small turtles would have faced predation by raccoons, muskrats, and raptors.

The spotted turtle hatchling I watch makes for the shallows of the marsh, dives, and finds the water a natural habitat, the seasons dictating its life. It keeps to shallow water and moves with extreme caution to avoid herons, muskrats, or large fish.

Some turtle hatchlings will winter in this aquatic domain; others will seek dry land and the comfort of leaves. Adults look for a safe place in the mucky water and settle into months of stillness, awaiting the warming sun.

6 September | MORE ON DRAGONFLIES

In his manuscripts, Thoreau referred to dragonflies as "devil's needles." They pirouette on sparkling wings, in plain view, but they are among the most elusive of insects. They hover, fly backward and sideways, mate while airborne, and turn 90 degrees on an aerial penny. Myths and folklore surround them, and one might ask, why the name "dragonfly"? It may stem from the Romanian word *drac* (meaning both "devil" and "dragon"), and it is surmised that St. George's horse, turned into a flying insect under the control of the devil, could have ended up as a dragon.

Seven of the 11 families of dragonflies occur in the United States. They are largely named according to their appearance and flight:

Skimmers: some can stay airborne for hours, but the majority perch in obvi-
ous sight (this family of dragonflies is called "perchers" in Europe)
Clubtails: have three enlarged abdominal segments, and they tend to be wary
Darners: known as hawkers in Europe, are strong flyers and fierce predators

Emeralds: possess striking emerald green eyes, are secretive, have brief flight
    seasons, and fly for only short periods
Cruisers: seldom are at rest as they patrol the reaches of rivers or streams
Petaltails: have petal-shaped appendages on the abdominal tips, and their
    cryptic coloration allows them to blend in with their habitat

I look at a skimmer, a common whitetail, that is stationary on a boardwalk rail.
With a chalk-white, plump abdomen and broad dark wing bands, it is spectacular
and easy to identify. It enjoys a long season, from early spring into autumn, with
flights that are fairly short and frequently interrupted by rest stops. The female
lays her eggs very speedily, expelling nearly 2,000 eggs per minute as she taps her
abdomen on the water's surface. In contrast, the smaller ruby meadowhawk lays
fewer than 40 eggs per minute.

Because common whitetails mate so fast, the male is often able to take a second
mate and then guard two females at once. Indeed, for the most part, dragonfly sex
is rough and tumble, and though females may mate with multiple partners, the
sperm from her final mating will fertilize the eggs. Consequently, a male dragonfly
maximizes his chances of fatherhood by destroying the sperm of his competitors,
either by scooping out previous inseminations or by adding ejaculate over that
present in a female's sperm storage organ.

Like many birds, some dragonflies are migratory, arriving early in spring from
more southerly wintering grounds. The adults breed, the young hatch, and the next
generation heads south in the autumn. Most dragonflies are residents, spending
most of their lives as nymphs, living underwater. We only see them as they emerge
and assume a winged form.

All this passing through my mind teases out Dante Gabriel Rossetti's poem,
"The House of Life: 19. Silent Noon," set for voice and piano by Ralph Vaughan
Williams:

Deep in the sun-searched growths the dragon-fly
Hangs like a blue thread loosened from the sky:—
So this wing'd hour is dropt to us from above.
Oh! clasp we to our hearts, for deathless dower,
This close-companioned inarticulate hour
When twofold silence was the song of love.

The Value of Streaming Youngsters

With D-shaped nets, sieve buckets, hand lenses (small, hand-held magnifying devices), ice cube trays, and plastic spoons, stream teams undertake regular samplings of Two Run Branch for benthic macroinvertebrates (or "stream bugs," as some would have it). The teams turn over submerged logs, leaf packs, and stones for larval dragonflies, caddisflies, mayflies, and other bugs that come their way.

Since 1996, quarterly studies (mentioned in many issues of *Marsh Notes*) have been conducted on the creek to monitor the diversity and abundance of stream insects, which are indicators of the health of their home, including the water quality. Many land use decisions impact smaller streams and creeks and, by looking at the watershed in its totality, more-informed land use decisions can be made in protecting the Patuxent River and, ultimately, Chesapeake Bay.

Increased or decreased water flow, soil erosion, the loss of riparian vegetation, fertilizers, and pesticides change the ecosystems of streams and rivers. To measure the impacts, the team examines benthic macroinvertebrates as an index of biological integrity, a scientific tool developed by Maryland's Department of Natural Resources to identify and classify water pollution problems. Some insects, such as mayflies and stoneflies, are very sensitive to pollution and survive in only the healthiest of streams. Others, such as black flies and mosquitoes, tolerate poor water quality.

To evaluate what's been happening at Two Run Branch, three measurements have been used: the total number of insect families; the number of mayfly, stonefly, and caddisfly taxa; and the percentage of mayflies. Totals for each of the taxonomic families were shown to vary with the seasons, with winter samples generally having the highest number of families. Cooler water temperatures and higher dissolved oxygen levels enable more-sensitive insects, such as stoneflies, to survive. In con-

7 *September* | STRING OF PEARLS

"A mosaic of preserved lands (pearls) and natural corridors (strings) holds the thin tissue of the planet together." So declares a speaker at a ceremony held at the Sanctuary today, honoring six farm owners for saving land "forever" and

trast, warm summer days result in higher water temperatures and lower amounts of dissolved oxygen. Equally, most insects will have left the stream to fly and propagate. So, taxonomic richness at the family level has remained fairly stable from year to year.

The number of mayfly, stonefly, and caddisfly taxa are found to be high in better-quality streams, and Two Run Branch rates as "fair" or "good" in 70 percent of the samples. Removing summer data, as the taxa were affected by drought, brings the amount up to 88 percent. Based solely on the percentage of mayflies, the creek was rated rated "fair" to "good" in 95 percent of the samples.

Natural variation in weather conditions has a great influence on how quickly insects mature and demonstrates the importance of factoring in variables other than the index of biological integrity. An accurate evaluation of stream quality depends on combining multiple metrics (standards of measurement) and dates.

The students on the teams were reminded that Two Run Branch empties into a floodplain—the haunt of birdlife favoring moist, shady environs. In summertime, the soaring maples and tuliptrees provide an unbroken canopy for foraging warblers and other insect-eating birds. In the lower canopy, northern cardinals, wrens, sparrows, and eastern towhees go about their business; in the floodplain, a variety of foods support Louisiana waterthrushes, common yellowthroats, hooded warblers, worm-eating warblers, northern parulas, and white-eyed and red-eyed vireos. Louisiana waterthrushes build their nests in the fern-covered banks sloping into the creek. Downstream, in the Beaver Pond, wood ducks, great blue herons, and belted kingfishers delight the observant birder, who may even be lucky enough to spot an occasional red-headed woodpecker. For the herpetologist, there are 5 species of turtles and 13 species of frogs and toads, dominated by green frogs.

encouraging others to do the same in the District of Columbia and the six states within the Chesapeake Bay watershed. Natural areas and farmlands connected by corridors for wildlife keep an increasingly fragile landscape intact, serving as the building blocks of natural systems that help maintain a healthy planet.

Through conservation easements, these farm owners have shunned development and adopted long-term ecological values. They either manage or oversee crop rotations and pastures, subject woodlands to forest stewardship plans, and conserve riparian vegetation and waterways. Many owners view preservation of their lands as a tribute to their parents and wish to play a small part in protecting a piece of the Earth for future generations.

Maryland has a growing history of land conservation and other efforts to protect its remaining critical habitats. In the 1970s, the Department of Natural Resources initiated the Upland Natural Areas Survey, and Maryland's Natural Heritage Program (now the Wildlife and Heritage Service) was formed in 1979 to identify the most ecologically important lands in the state and protect them for rare species and important natural communities. Further work produced Maryland's Green Infrastructure, a mapped network of large blocks of intact forest and wetlands (called hubs), linked together by linear features (called corridors, or habitat highways), such as forested stream valleys, ridgelines, or other natural areas.

The Green Infrastructure approach is based on principles of landscape ecology and conservation biology. It has created an ecological network map (GreenPrint), using satellite imagery to characterize land cover; geographic information system data on streams, wetlands and other resource features; and biological databases. With these pieces of information, it has identified areas as potential candidates for restoration activities.

Land conservation programs have been enacted, including agricultural preservation, private conservation easements, and regulations preserving wetlands and shorelines. They are piecemeal approaches, however, and much of Maryland's green infrastructure is currently unprotected. Consequently, there needs to be the will to change and to provide investments in conserving and rebuilding natural assets on a landscape scale to capture and sustain ecological integrity.

10 September | BEAVERS: SECOND ONLY TO HUMANS IN CHANGING THEIR ENVIRONMENT?

I take a look at the dam repaired by North American beavers in the Otter Pond—now flooding much of the pond—with the usual suspects enjoying their reclaimed habitat: turtles sunning on logs, watersnakes sneaking up on prey, and belted kingfishers diving after small fish. The beaver's dome-like home has been constructed in the center of the impoundment and may only be reached by underwater entrances. Not until times like this, in the dying light, do these twilight and nocturnal creatures venture forth. I sit downwind to prevent my scent or any noise from reaching them, and I eventually see a flattened tail break the waterline, acting as a rudder and moving its owner to the shore. A beaver's waddle on land is in marked contrast to its graceful progress in the water.

Behind me, tree stumps show gnaw marks from beavers: on mature trees employed for dam construction, and on young trees used for food. The beavers usually chew the trunks and branches into small sections before stripping and eating the inner bark.[4] Regrowth in subsequent years will offer these mammals an easy reach to stems and leaves. In the pond, drowned trees provide both standing and fallen wood, creating habitat for many plants and animals.

12 September | SOIL: THE FINAL FRONTIER?

As I walk the trails of the Sanctuary, crossing grassland, woodland, and marsh, I often wonder about the soils on which I tread. Whether it is sandy, loamy, or hydric (soggy and mucky) in nature, soil is an ecosystem unto itself teaming with life. In this part of Maryland, the soil is also a reflection of the underlying geology from the Tertiary era. Plant roots probe the mix of minerals and organic matter, the latter composed of material digested by bacteria, fungi, protozoa, nematodes, earthworms, arthropods (insects, crustaceans, and spiders), and other forms of life. Such biodiversity is dependent on plants, since they exude proteins and carbohydrates, forming a narrow region (a rhizosphere), without which most nutrients would be leached from the soil.

It is estimated that 1 billion bacteria inhabit a single gram of productive soil (representing from 50,000 to over 1 million species)—equivalent to a ton of bacteria per acre. Few of these bacterial species have been described, and we know less about life in the soil than we do about the far side of the moon. It is a world where fungi trap thread-like worms, and some bacteria consume toxic chemicals.

Many bacteria are decomposers, converting energy into forms that are useful to organisms in the soil's food web. Nitrogen and other elements essential to plants are stored in decomposer cells, preventing their loss from the root zone. When the bacteria die, plants may absorb the stored nutrients.

Saprophytic fungi are also important decomposers, breaking down dead or dying organic matter into humus, minerals, and nutrients that plants can absorb.

Other fungi, called mycorrhizae, colonize plant roots and sustain a mutually ben-eficial relationship by increasing the root surface's absorbing area by 100 times or more. In return for the plants' carbon-containing exudates, mycorrhizae provide them with water and nutrients.

As they burrow, earthworms, ants, termites, and moles, along with some mil-lipedes and beetles, churn the soil and influence its continued formation. Earth-worms ingest soil particles and organic residues that, when voided, enhance the availability of nutrients for plants. They also aerate and stir the soil, improving the infiltration of water.

Plants help reduce erosion from excessive rainfall and surface water runoff. They also shade soils, keeping them cooler and slowing the evaporation of mois-ture. Plants generally lessen soil moisture variations, preventing soil from desic-cation during the driest months and drying it during moister months. They can form chemicals used by mycorrhizal fungi and rhizospheric bacteria to break down minerals and, thereby, help improve soil structure.

Whereas most soils are aerobic, allowing plants to consume oxygen and nu-trients while releasing carbon dioxide, wetland soils are anaerobic, resulting in intense competition by soil organisms for the limited oxygen. In such hydric soils, different chemical and biological reactions dominate, necessitating various kinds of survival strategies by occupying organisms. Wetland plants, such as cattails, sedges, and waterlilies, use aerenchyma (spaces in their stems and rhizomes) to transport atmospheric oxygen to their roots. Microorganisms degrade organic matter more slowly in anaerobic soils than in aerobic ones. Consequently, this material accumulates at the surface and is often referred to as peat or muck.

Wildlife benefit from fallen leaves. Frogs, turtles, and salamanders use them for concealment and hibernation, and the caterpillars of many moth and butterfly species overwinter in leaf cover before emerging in the spring. Earthworms and millipedes help decompose leaf litter by returning minerals and nutrients to the soil. Such occupants, in turn, are a rich food source for toads and birds.

Soil is an important—and often-neglected—element of climates. After the oceans, it is the second largest place for carbon storage, or sequestration. A changing climate can impact nutrient cycling, ecosystem respiration, and carbon storage in forests. Depending on the region, this may result either in more carbon being stored in plants and soil, due to increased vegetation growth, or more being released into the atmosphere.

While global models predict that climate change can increase global net primary production, regional variations in climate, nutrient availability, and the amount of available water will have the largest impact on tree growth locally.[5]

One last point often made by naturalists—dirt is what you find under your fingernails; soil is what you find under your feet.

15 *September* | Confusing Fall Warblers

What makes the marvelous is its peculiar way of being ordinary; what makes the ordinary is its peculiar way of being marvelous.

Orhan Pamuk, *The Black Book*

At this time of year, two sets of birdwatchers emerge: those moaning as they struggle to identify fall warblers, and those delighting in the challenge. In the spring, the warblers arrive with bright, colorful plumages (making them the "butterflies of the bird world") and distinctive songs. In autumn, they sport dull, nonbreeding plumage, as do their similarly colored juveniles, and depart in silence.

Not surprisingly, there is an industry available to offer help in the form of books, websites, and instruction by experienced "warblerites." All of these sources note that while color is important, certain markings, locations, and behaviors also provide reliable clues.

In his classic 1934 book, *A Field Guide to the Birds*, Roger Tory Peterson laid out two full-color plates of "confusing fall warblers." He employed the innovative technique of highlighting key field marks to differentiate the various species. Though intended to help, it caused many beginning birders to abandon their binoculars in frustration.

Nonetheless, with some work and patience, confusion may be contained. An especially helpful publication is the *Warbler Guide*, by Tom Stephenson and Scott Whittle, denoting the 56 species of warblers in the United States and Canada. It has become a groundbreaking guide, featuring more than 1,000 color photographs, species accounts with multiple viewing angles, and a new system of vocalization analysis to help distinguish songs and calls. It is also available as an app.

When their guide came out in 2013, Tom and Scott gave a presentation on it at the Sanctuary. They filled the Center and held the audience in thrall. Many swore that their lives were changed, and the guide has since become a bible for birders.

I must confess that a walk with experienced birders still gives me the most valuable insights. They spend time on identifying a bird, teasing out the essentials (including behavior and location), and I take advantage of this during today's bird count around the Sanctuary. We see a Nashville warbler (male: gray head, bold white eye ring, and short tail, which it bobs and swishes left and right), a common yellowthroat (female: yellow throat and breast, white belly, wren-like cocked or flicked tail, no mask), a hooded warbler (male: black hood surrounding the face, olive above, yellow below, frequently flicks open its tail), and a black-throated blue warbler (female: dark cheek, white wing spot, holds its wings out slightly while foraging). There were some immature species in the treetops that were subject to much analysis but ultimately defied identification.

As these bundles of feathers depart for neotropical climes, I am drawn to the problems they face. Development is gobbling up their summer and winter habitats. Skyscrapers, other high-rise structures, and communication towers kill many millions every year. Take, for example, cerulean warblers. The North American Breeding Bird Survey estimates a decline of over 2.6 percent per year between 1966 and 2015, resulting in a cumulative loss of 74 percent of the population of these sky blue birds. These warblers rate a 15 out of 20 on the survey's "Continental Concern" score and are on the 2016 "State of North America's Birds Watch List" of species considered most at risk of extinction without significant conservation actions to reverse their declines and reduce threats to them. In Maryland, cerulean warblers breed in the Cumberland Plateau, but as the *Second Atlas of the Breeding Birds of Maryland and the District of Columbia* states, "the 40 percent net decline in between atlas projects in the 1980s and 2000s is a clear warning that this warbler requires careful monitoring, habitat preservation, and management if it is to remain in Maryland's and DC's avifauna."

Brown-headed cowbirds, which lay their eggs in the cerulean warblers' nests, may be finding their unwitting hosts more easily as forest patches become smaller.

At night the calls of migrants fill the sky as the birds pass high overhead, sometimes competing with the fading, if persistent, singing of crickets and katydids.

16 September | OSPREYS LEAVE

The plaintive cries of ospreys no longer echo across the marsh. Usually alone, though sometimes migrating with other raptors, individuals in eastern populations are heading for Caribbean islands or South America. Their average speeds of travel are 126 miles per day, and, unlike other falcons, ospreys undertake extensive water crossings, facilitated by their sizable yet narrow wing length. Females invariably precede males, leaving before their young are completely independent. Mates will not see each other again until they return to their nests in the spring.

Studies find no difference in their fall migratory routes by sex, although females tend to winter farther south than males from the same breeding population. Juvenile ospreys do not leave their wintering grounds until their second spring, and many stay until their third spring.

An osprey may log more than 160,000 migration miles during its 15- to 20-year lifetime. During 13 days in 2008, one osprey flew 2,700 miles—from Martha's Vineyard in Massachusetts to French Guiana in South America. In a 17-year study satellite-tracking 118 ospreys, Mark Martell and colleagues have provided much-needed data revealing migrational differences among ospreys and have helped define strategies to deal with threats to the species.

19 September | THE WONDERS OF MONARCH BUTTERFLIES

The flitting brilliance of orange to reddish wings, alternating with prolonged glides, now fills many a bush and, against today's blue skies, arrests time and motion. They are monarch butterflies on their southerly migration—sometimes pausing by the thousands on bushes and trees—and provide one of nature's true spectacles.[6]

Sustained by fat reserves and periodic feeding (having gone through three generations during the summer), monarchs will travel several thousand miles in soaring flights. They normally cover 50 miles a day, but can fly up to 80, taking about two months to reach their overwintering sites in the fir-clad mountains of central Mexico. One site has been reported to support about 12–15 million individuals on just three to four acres. In early spring, the adults move north to lay eggs, with their offspring then continuing north later, reaching eastern and central states in April or May.

As fall approaches, an exceptional generation of butterflies is born. Unlike their parents, grandparents, great grandparents, and great-great grandparents, whose lives are counted in weeks, this migratory generation survives for seven or eight months, truly a Methuselah generation. As the sun keeps moving, monarch butterflies constantly adjust to stay on course throughout the day. Like birds, winds and topography affect monarchs, and it is only recently that scientists have determined how they navigate and locate their Mexican winter roosts. Christine Merlin and colleagues have found that their antennae are key, for, when these appendages were dipped in colored enamel paint, the affected individuals were unable to orient. In contrast, butterflies whose antennae were coated with clear paint had no trouble navigating.

This year, monarch butterflies got a late start to their migration, and scientists fear that climate change is behind what they are calling the tardiest monarch migration ever recorded in the eastern United States. The scientists also worry that rising temperatures pose a new threat to a species whose numbers have seen record lows in recent years.

21 September | AUTUMNAL EQUINOX

Today, at 4:04 pm, the sun's winter retreat to the Southern Hemisphere marks the autumnal equinox. Waning sunlight will increasingly impair the ability of vegetation to photosynthesize and will begin to force changes in its color. It's a case of dwindling chlorophyll activity and declining nutrients, so leaves and other forms of vegetation drop their burden to carpet woodland floors or form mats on marsh mudflats.

Some trees may shed leaves during midsummer, when temperatures are high and the bulk of growth has ceased. For most, however, it occurs in the fall. Seasonal changes, including chemical processes, provoke this phenomenon, with a shedding zone at the base of the leaf, closing off the stem. Some organic compounds, such as nitrogen and phosphorus, are recycled. As leaves gradually dry out, they are wrested off by their own weight, as well as by rain and wind.

But why the yellows, oranges, and reds in place of the disappearing green? One suggestion is that the red-based hues help protect the dying leaves from damage

"Fall" versus "Autumn"

Thinking of the year in terms of four seasons has been an evolving concept. Anglo-Saxons marked the passage of time with but one season—winter, derived from a root word meaning "wet" or "water"—reflecting the hardships typical of their entire year.

Summer is also time bound. In Old English, the word *gear* connoted the warmer part of the year, eventually yielding to the Germanic *sommer* or *sumar*, related to the word for "half." Eventually, speakers of Middle English (the language used from the eleventh to the fifteenth centuries) thought of the year in terms of the warm half—"summer"—and the cold half—"winter."

In twelfth- and thirteenth-century Middle English, spring was known as "lent" or "lenten," and fall as "haerfest." In the fourteenth and fifteenth centuries, "lenten" ceded to "spring," "spryngyng tyme," "ver" (Latin for "green"), "primetemps" (French for "new time"), and more-complicated descriptive phrases. By the seventeenth century, "spring" triumphed.

Fall obtains its name from the longer phrases "fall of the year" or "fall of the leaf" ("spring" arises from a similar phrase, "spring of the leaf"). "Autumn" came to English from the French *automne* in the 1300s, but it did not gain prominence until the eighteenth century. *Autumn*, a Latin word, first appears in English in the late fourteenth century. In the seventeenth century, "fall" came into use, almost certainly as a poetic complement to "spring." Finally, in the eighteenth century, "harvest" lost its seasonal meaning, and "fall" and "autumn" emerged as the two accepted names for the year's third season.

Even when people spoke of fall or autumn, they couldn't agree on precisely when it took place. In the seventeenth and eighteenth centuries, dictionaries by Thomas Blount and Samuel Johnson noted that some thought fall began in August and ended in November, while others contended that it began in September (at the equinox) and ended in December (with the solstice).

Later, while "fall" became the preferred term in the United States, "autumn" became prevalent in British English. This has changed, however, as "fall" has been gaining ground in British publications for some time.

Source: Wolchover 2012.

by the sun, thereby allowing them to collect the sun's energy a trifle longer. The red coloration may also discourage animals from eating the leaves or laying eggs on them. In time, sunlight and frost destroy all remaining pigments, save for the brown tannin, and put paid to any surviving cells, rendering the leaves dry and crunchy underfoot.

Trees bathed in sunlight tend to produce more reds, and trees in the shade, more yellows. Temperatures below 45°F but above freezing foster reds and purples. Yellows and browns always appear at any autumnal temperature. In general, warm sunny days with cool nights create the optimum conditions for the best fall color. This year, because of unusually dry weather in August and September, the colors are among the weakest in seven years. Some areas of Western Maryland, however, are having a good year.

23 *September* | And What Fall, or Autumn?

Yesterday, record heat scorched the Upper Midwest and Great Lakes and is now blanketing the Mid-Atlantic and Northeast. In the Washington, DC, region, temperatures have reached the 90s and are expected to continue until midweek, some 10–12 degrees above the norm. At the week's end, a big cold front is expected to reverse the pattern—abruptly bringing in more fall-like weather.

Tiny midges fill the air, having just hatched from the marsh. Although they resemble mosquitoes, these nonbiting insects survive only long enough to mate. Eggs are laid in masses in the water, and several generations are bred during the summer. The current dry weather may well have extended their breeding season.

After hatching from eggs, the emergent midge larvae burrow into the mud or construct small tubes in which they live. As they grow, the larvae take on a pink color, gradually turning a dark red to become the familiar bloodworms. The larval stage can last from two to seven weeks (depending on water temperature). Larvae eventually rise to the surface as pupae, attaching one end to the surface film, with the rest of the body dangling vertically below.[7] Adults emerge several hours later, forming the swarms I see before me.

In the fall, mature midge larvae survive the winter months by suspending their development. Pupation and the emergence of adults both occur the following spring, in late March or early April. Several more generations of midges will be produced throughout the summer, resulting in continuing mass emergences of adults.

Midges are beneficial, as they provide food for a variety of freshwater fish and other aquatic creatures. These little insects also helped the Cleveland Indians win the 2007 American League Championship Series when a swarm unnerved New York Yankees pitcher Joba Chamberlin.

27 *September* | SORAS, OR MASKED MARSH CHICKENS

In a shallow boat, Greg Kearns stops at the entrance of a partially netted corridor he built through the cattails, in order to capture soras newly arrived in the Jug Bay marsh. We clamber over pallets, holding on to the cattails, and, rounding a corner, see a trap with a captured sora, or sora rail. Because of the rising tide, Greg's interns have to be quick to seize the bird and bring it to the boat for banding.

What a delightful bird it is: small and plump, gray-brown, short, with a cocked

A sora amid wild rice. *M-NCPPC Department of Parks and Recreation*

Soras and Wild Rice

In 1987, Greg Kearns, a naturalist at Patuxent River Park, began a study of soras using Jug Bay's marshes to learn more about their migration patterns, local movements, and mortality rates. Though seemingly weak flyers, they journey south to the Gulf of Mexico and South America from northern marshes between August and November, using the marshes of Jug Bay as a resting and refueling area. They stay up to 45 days, consuming wild rice and halberdleaf tearthumb seeds before the first frosts. More often heard than seen, their distinctive calls provide clues to their identification and location. Their long toes ease these birds' passage over soft mud, while their stout bills grab seeds and insects. They take readily to water, swimming on the surface and sometimes using their wings to propel themselves underwater.

As wild rice populations began to experience marked declines, a report by Brooke Meanley in the fall of 1999 questioned the reference to Jug Bay as the "Patuxent River wild rice marsh." The then extant stands of wild rice represented only 10 percent of those existing 30 to 40 years in the past, where they dominated the marsh south of the Railroad Bed. Areas once beautiful and plentiful gave way to barren mud or sad remnants of short grazed plants. To quote Greg Kearns, in an article by Elaine Friebele: "In the fall, there were dozens and dozens of soras in the Wild Rice [near the Observation Blind]. Often 50 would be calling at once. Now, in the 1990s, there's little Wild Rice, and it's difficult to hear Soras there." In parentheses, Greg adds that in the fall of 1960, 2 million red-winged blackbirds

tail, buffy white undertail coverts, and a yellow bill. The absence of a dark throat patch marks it as an immature bird. Greg and his interns measure the bird's beak, large leg bone, and middle claw. With a final weighing, they determine that it is a first year female. I handle the sora and am told to carefully compress her very flexible ribcage, so I can fully appreciate how this adaption allows her to negotiate dense plant growth.[8] When Greg gently unfolds his hands, the bird rises and looks around, seemingly in no hurry to leave. When she finally departs, she scurries across the river, with legs dangling, and drops into the cattails, as if unable to fly farther.

and 10,000 bobolinks (reedbirds), as well as many soras, fed on ripening wild rice and other seeds in the marshes.

Some specialists held that sedimentation was largely responsible for this decline, while others believed the grazing of seedlings was the main cause. Greg (and others) supported the latter view and set about constructing wire enclosures around seedlings in late April. By July, tall, dense stands grew within the enclosures, while the surrounding mud remained either barren or covered in very sparse wild rice growth. It turned out that the major culprits involved were various introduced Asian carp species and resident Canada geese, though Greg thinks it may be that a rising sea level, heightened sedimentation, and nutrient deficiencies could also affect the growth rate and health of the seedlings.

Sedimentation and tidal washing affect plant distributions in the marsh, with deposits along the edge resulting in the formation of a low levee, some two to three feet wide. Barnyard grass grows at the higher levels and is an important food for soras and some ducks. During high tide, soras gather in the grass and the accompanying dotted smartweeds—a behavior known to hunters. Lower areas are usually composed of halberdleaf tearthumbs and cattails, which are invariably replaced by rice cutgrass toward the shore and pure stands of wild rice at lower levels.

Red-winged blackbirds still come in goodly numbers, but bobolinks are a rarity. As they move from one wild rice plant to another, the red-wings scatter some of the seed, which is picked up by the soras, blue-winged teal, American black ducks, and wood ducks. Later, when the wild rice plants fall over, soras and teal can easily reach the grain heads.

On the open water, Greg slaps the surface with a paddle to elicit responses from soras. The startled birds answer with sharp "keek." Current wind patterns have held back their migration, and as many as five to eight respond at any one time.

Moving to other traps, we pass ever-browning cattails, dotted smartweeds, water millets, golden bands of burr marigolds, and patches of purplestem (or marsh) asters. This will be the last splash of color before spring returns, and the marsh now pulsates with the newly arrived northwesterly winds.

A skiff with a pusher (*center*) and two hunters harvesting soras. *M-NCPPC Department of Parks and Recreation*

In our last visited trap, a lone female sora has been caught, along with two marsh rice rats. The latter take some careful handling, for their teeth can penetrate many a glove and impart a painful bite. These thick-furred, grayish-brown creatures are barely five inches in length, including their tails. With off-white feet and underparts, they are quite exquisite. Northern harriers passing through these marshes must find marsh rice rats to be good eating, but the birds have to be diligent, for these rats are largely nocturnal.

30 September | CLIMATE: WARM, WEIRD, AND SPOOKY

These last few days have been sunny but crisp, with blustery northwesterly winds backed up behind a cold front, bringing temperatures back to normal: low 70s during the day and mid-50s at night. Predictions are that next week will see temperatures return to the mid-70s and near 80, with winds streaming from the southeast.

Dry conditions continue, with precipitation some two inches below normal. Soil moisture is low, but an official drought has yet to be declared. Some forecast-

The Marsh and Sora Hunting

Soras (sora rails) have been hunted by the tobacco aristocracy since Colonial times, but the advent of the railroad opened the Jug Bay area to more hunters, especially from the wealthy class. In the early 1900s, it attracted the likes of Teddy Roosevelt, Babe Ruth, Harry Truman, and General Billy Mitchell. The hunters would gather at the Pig Point docks just before high tide, hire a pusher, and sail out in small skiffs (rail boats).

The hunt proved to be hard work for the pusher, who would stand at the stern and maneuver the boat through the marsh with a 14-foot pole bearing three cleats on the bottom, to prevent it being seized by the mud. Practiced pushers positioned the rail boat quickly, to provide hunters with the best chance of a shot at the awkward-flying soras. They also picked up the dead birds, recalling where they had fallen.

The closure of the railroad in 1935, denying easy access for hunters, was accompanied by a general decline in sora hunting. Natural factors also played a part: farming and development increased siltation and made the marsh unnavigable by rail boats. Above all, the soras, having been overhunted, were no longer plentiful.

ers have warned of "summer's revenge, part two," with a heat dome building over the central and eastern parts of the nation and computer models predicting that October will be warmer and drier than normal—a lot like this September. The first half of the month has been unseasonably cool and dampish, the second abnormally warm and dry as dust. October, however, is notorious for big temperature swings, so another roller coaster may be in store.

Guy Walton, a retired meteorologist from the Weather Channel, has described the weather pattern as " warm, weird, and spooky."

Whatever is coming, the northeasterly winds have unblocked stagnant air masses and pushed oodles of birds through the skies. The passage of monarch butterflies had steadily increased these past few days, but they are clustering in roosts today until the wind strength abates.

September is the forerunner of the gorgeous tapestry that will envelop trees in the coming weeks. There is also a last burst of flowers, which drew the attention of William Cullen Bryant in his poem, "To the Fringed Gentian":

Thou waitest late and com'st alone,
When woods are bare and birds are flown,
And frosts and shortening days portend
The aged year is near his end.

Then doth thy sweet and quiet eye
Look through its fringes to the sky,
Blue-blue-as if that sky let fall
A flower from its cerulean wall.

Migrant birds now head south at full throttle, and the air thickens with swirling flocks before they settle on trees and shrubs and along marshes and rivers. The thin blue skies of early morning contend with mist, and the final wave of emergent insects and migrant butterflies and moths sup their last drops of nectar. Winter visitors begin to appear. With a bonanza of seeds and fruit, squirrels and other animals prepare for hibernation or store bodily fat for the harsh conditions ahead. Warm days cede to October's cold nights and the evening appearance of owls. Nature is entering its resting period.

Edge effect from grassland to woodland in autumn. *Frode Jacobsen*

OCTOBER
The Leaf-Falling Moon

I have been younger in October
than in all the months of spring.

W. S. Merwin, "The Love of October"

1 October | FISHING SPIDERS

Six-spotted fishing spiders (raft spiders), with a striking pale stripe down each side of the body, are poised at the watery edge of the marsh to detect ripples from prey and then scamper across the surface of the water to subdue it. Using small claws in their foremost legs, they inject venom to kill and then help them digest their mainly aquatic insect prey. Sometimes they capture tadpoles or small fish.

These spiders are covered in short, water-repellent, velvety hairs, allowing them to use the water's surface tension to stand or run on it, just like water striders (pond skaters). The six-spotted fishing spiders can also move beneath the water, with air trapped in the hairs forming a thin film over their bodies and legs, giving them the appearance of finely polished silver. Like other spiders, they have a series of very thin, leaf-like structures (book lungs) beneath the abdomen, open to the air film on their bodies, thereby allowing them to breathe while submerged.

Six-spotted fishing spiders are nocturnal hunters. Holding on to the shore with their back legs, they stretch their front legs out on the water and wait for vibrations carried by it, just as other spiders feel vibrations in a web. They can distinguish the drawn-out, erratic vibrations of a struggling insect from the one-off vibrations caused by falling leaves, the background noise of the wind, or the flow of the water around rocks and other obstacles. As well as identifying the source of the vibrations, these spiders can discern the distance to and direction of the source. To do this, they use very sensitive hairs on their legs and feet. Their eyes play a secondary role, being of less use for nocturnal hunting.

A six-spotted fishing spider. *Frode Jacobsen*

As soon as the vibrations reveal a floundering insect within range, fishing spiders may take direct action, running across the surface and grabbing the insect before it extracts itself from the water and flies to safety. Some fishing spiders use silk draglines to prevent themselves from speeding past their prey, which are mainly aquatic insects. The spiders can dive up to seven inches underwater to capture prey. In this instance, their good vision contributes to their success.

2 October | WASPS: SOCIAL AND OTHERWISE

On a princely afternoon, my wife and I have walked through the woodland and followed the tracks alongside the marsh, and we now stand before a meadow alive with the droning of wasps. Many of the accompanying bees, flies, and butterflies are potential prey, but today the wasps are exploring the remaining flowers, in hope of extracting the last traces of pollen.

We are searching in particular for social wasps (often called yellowjackets), familiar to many, if only for their sting. They live in colonies composed of hundreds or thousands of sterile female workers and their egg-laying queen. Their nests, started in late spring (as the queens emerge from hibernation), are constructed from wood fibers chewed by the industrious workers to make a kind of papier maché.

Wasp workers toil endlessly to raise sister workers from eggs laid by the queen, cooperate to build and defend the nest, collect food, and care for the queen. When the colony reaches a certain size, the workers feed some young larvae larger amounts of food, stimulating the development of a potential queen. After mating with males from other colonies, these queens overwinter before starting a colony of their own the following spring. Male wasps, which take no part in the social life of the colony, develop from unfertilized eggs laid by the queen and, rarely, by workers.

Wasps—despite leading these intriguing lives, playing a major role in the natural environment, and facing problems similar to those of bees—are always in need of a public relations campaign, being excoriated by many. As to their ecological significance, social wasps are predators and are vital in controlling the numbers of potential pests: ticks, houseflies, blowflies, greenflies, and many caterpillars and larvae. For example, so great are the food demands of growing yellowjacket colonies that more than two pounds of insects may be removed from a 2,000-square-foot garden. Wasps are also increasingly recognized as valuable pollinators, transferring pollen as they visit flowers to gather nectar.

Another group of wasps deserving of our attention are the solitary (nonsocial) cuckoo wasps, or emerald wasps, known for laying their eggs in host nests, where their larvae consume the host's eggs or larvae, as well as the food provided by the host. Cuckoo wasps are divided into two types, based on their lifestyles: parasitoids and kleptoparasites. Parasitoids feed on the larvae of the host, while kleptoparasites steal the host's food—behavior resembling that of common (or European) cuckoos. In both cases, the host larvae die. Hosts of parasitoid species include bees, sphecid (thread-waisted) wasps, potter wasps, sawflies, silk moths, and the eggs of stick insects. Kleptoparasitic species feed on provisions stored in sphecid wasp nests, which may include dead spiders, true bugs, aphids, or thrips.

If a cuckoo wasp parasitizing ground-nesting bees and wasps succeeds in depositing its eggs, the larvae have two survival strategies. Some eat both the host's larvae and their food items immediately; others wait until the host larvae finish their food supply and reach full size before consuming the host larvae. The first option requires the cuckoo wasp to eat several different kinds of food before it can pupate, while the second strategy lets the host larvae do all the work, eventually converting food stores into one juicy meal.

Pondering these and other marvels, we steal away from the carnival of color with which trees and shrubs are bidding their long farewell to summer.

4 *October* | THE SPIRIT OF SYNTHESIS: DIVERSITY OF RESEARCHERS AND WIDER APPRECIATION

Ecologists emphasize the importance of biological diversity at ecosystem, species, and genetic levels and support this mantra through international agreements, policies, regulations, and education. It is only when scientists come together with a particular perspective and a setting with which they identify, however, that they fully appreciate the advantages of providing their own diversity. For example, to protect a specific species and its habitat, they need to ensure variety in conducting aspects of their research. This diversity includes many factors: locations and their features, weather conditions, life stages to be studied, techniques, and even differences in researchers' ages and professions, as well as covering a range of observer biases or preferences. These, perhaps, are helpful in the same way divergences in techniques are useful.

Much as this is accepted by many practitioners, I often come across a blinkered appreciation of the natural world and a denial of intellectual insights and nourishing moments. Along with birders, butterfly watchers celebrate the fall of the year, with its promise of strays and vagrants. Both parties frequently share the same habitats, but it is surprising how little curiosity each group has about the other's interest. I well recall an occasion when my wife and I were wading into a lantana bush on the edge of a forest to admire some butterflies and other insects. We were met by three birders asking "Anything of interest? We've just seen a parula warbler." "Yes," we replied, "we're looking at a red admiral butterfly." A squall of pain crossed their faces, and they quickly turned away in search of more compelling rewards—for them!

In his book *Consilience*, the eminent biologist E. O. Wilson observes that the compartmentalization of the sciences is an artifact of scholarship, not of nature. Our historical approaches and institutional arrangements are not capable of handling problems requiring communication across disciplines, such as environmental policy, which combines the biological and social sciences. These and allied fields are not united in a common language or a set of principles that would allow collaborative and multidisciplinary efforts. Happily, David Attenborough is spearheading an effort at Cambridge University to bring all disciplines—"not just other botanists, not just ecologists, but law, international lawyers, psychologists, geographers, political scientists, and so on"—to bear on the pressing problems of the planet.[1] "Break down those walls and get people talking about it who would not otherwise meet."

In *Living in the Anthropocene*, John Kress and Jeffrey Stine see opportunities to move forward with conservation and habitat restoration, provided that these two issues engender a cooperative spirit between "citizens, governments, social and religious institutions, and the private sector." As E. O. Wilson says in the book's afterword, we can save ourselves as a species only if we save other species by large-scale habitat protection.

6 October | PRESERVED BERRIES

Though we have had a dry summer and an early autumn, many bushes are laden with berries. American black elderberry bushes arching over the marsh and the railroad track are replete with clusters of small black berries, and the red-purple fleshy fruits, with the seed enclosed in a single hard shell of mapleleaf viburnums are evident in the upland forest. Common winterberries in swamp thickets sport scarlet berries, while those of arrowwood viburnums, on the edge of the marsh are dark blue.[2]

Some of these and other berries last into winter, and even spring. I find Bernd Heinrich's book *Winter World* offers a plausible explanation for why certain ones endure longer than others. He suggests that sweet, juicy berries are quickly gobbled up by birds heading south and by mammals fattening themselves for the winter. The remaining berries are described as "dry," "acidic," "bitter," or some combination of adjectives, and winter-resident birds try to avoid this apparently meager fare. Heinrich describes the plants that grow these holdovers as "stingy" with juice and sugar. And there is probably no stingier berry than that of a staghorn sumac, whose berries are so hairy and lean that they seem more likely to be spread by the wind than by the eat-and-disperse method.[3]

Many birds require berries to fatten up for their long fall migratory flights, some putting on as much as 10 percent of their body weight per day. Adjustments are made in gut length and other digestive adaptations to allow fruit to be processed rapidly. Equally, the fruit's nutritional content depends on the season for which its dispersal is geared. This complex relationship of bird adapted to berry and berry to bird is as intricate as that of flowers to bees.

A close-up of a common winterberry bush in autumn. *Kerry Wixted*

Could it be that the overall strategy of some shrubs is to avoid having their berries eaten by local residents in winter, so as to hitch a ride in the spring with the long-distance (and hungrier) migrants when these fruits provide the only buffet in town?

Seasonal berries are a significant food resource for small songbirds during their autumn migration, so the effective conservation and management of important stopover sites that are used by migrating birds presupposes knowledge about the dietary requirements of songbirds and the nutritional composition of commonly consumed fruits.

With this in mind, a decade ago Susan Smith and fellow scientists measured the nutrient composition and energy density of nine common fruits and conducted a field experiment to estimate the consumption rates of three of these fruits by birds during their autumn migration. Most of the common fruits primarily contained carbohydrates, with little protein and fat, although three contained more fat: northern bayberries; arrowwood viburnums, or southern arrowwoods; and Virginia creepers. The scientists estimated that migratory birds on the study site (Block Island, Rhode Island), must eat up to four times their body mass in fruit wet weight each day to satisfy their energy requirements when eating low-energy fruits, such as those of American pokeweeds, and they cannot satisfy their protein requirements when eating only certain high-energy fruits, such as those of arrowwood viburnums. The results suggest that many migratory birds must eat both fruits and insects to meet their dietary needs. Thus shrubland habitat at important migratory stopover sites should be managed so that it contains a variety of preferred fruit-bearing shrubs and an adequate abundance of insects.

8–16 October | WEATHER

We have had no rain for 24 days in a row, but that ended on 8 October, with remnants of Hurricane Nate pushing tropical moisture into the region. Temperatures have reached the upper 70s to the 80s (10–12 degrees above the norm), with fleas and other insects enjoying the warmth, the former reaching a record abundance this summer. The record for the most humid period for October, with 80°F weather, occurred at night on 9 October. The following two days have brought cooling temperatures and sporadic rain has fallen for the four days afterward. This is much needed, as September was some two inches short on precipitation.

Summer's swan song arrived yesterday (15 October), with temperatures close to 80 on a very humid afternoon. Today has brought the season's first frost advisories in some parts of the region, and the forecast for the next two weeks is for highs in the 60s and 70s, in line with fall weather averages.

Looking back, the year's summer-like conditions began on 9 June, and many declared it spring's end. While it proved to be a long summer, the weather was not particularly intense, aside from a stretch from late June to late July. By historical standards, August was quite pleasant, as temperatures averaged slightly cooler than normal.

But the summer did not relent. After the coolest start to September in 86 years, 16 out of 17 days hit at least 80°F, lasting from 12 to 28 September—including three days above 90°F. Then 10 of the first 15 days of October were 75°F or higher, including six straight days above 80°F, tied for the fifth-longest streak on record for October.

The summer of 2016, while more intense than this year's, was slightly shorter —beginning on 24 May and ending on 23 September. In 2015, summer spanned the period from 8 May to 11 September.

12 October | THE FATE OF THE EARTH, OR THE
BLUE MARBLE

Yesterday, Elizabeth Kolbert, a *New Yorker* magazine staff writer, delivered the second annual Jonathan Schell Memorial Lecture, an event established by the nonprofit Nation Institute in honor of the late Jonathan Schell, a longtime *New Yorker* staff writer. It is named for "The fate of the Earth," a series of articles that Schell wrote for the magazine in 1982, later published as a book.

Kolbert began by saying that October 2017 was "a scarily opportune moment to talk about nuclear war or to talk about climate change—or to talk about climate change *and* nuclear war." Nonetheless, she went on to state:

> I am going to try to do something different. Instead of looking at the fate of Earth from our anxious perspective, from a human perspective, I'd like to try to look at it from the viewpoint of the millions and millions of non-human species with which we share the planet. This represents a different kind of imaginative exercise. It requires us not to imagine events that might happen but to look at events that *have* happened through different eyes—or even without eyes, since so many of our fellow creatures lack them. . . .
>
> Today, the biomass of Earth's human population is estimated to be ten times greater than the combined biomass of all the planet's wild mammals. . . . Meanwhile, if we look at the weight of our domesticated animals—cows and goats and pigs—the situation is even more extreme. Their biomass is roughly twenty-five times greater than that of wild mammals. And if you add our beasts and us together the ratio is thirty-five to one. In numerical terms, we are a hugely successful species—an astonishingly successful species—and our success has come at the expense of other living things.

Kolbert ended by observing that in "October of 2017, it's easy to worry that the human project is in danger. From the perspective of other species, though, what's scary is not the fragility of human life but its remorseless vigor. We should attend to the fate of Earth for our own reasons. The greatest threats that we face—nuclear war, climate change—are almost easier to accomplish these days than they are to envision. But as important as we are to ourselves, we're not all there is on this blue marble. And if we are just thinking about ourselves, then we are failing as ethical agents, which is to say as human beings."

According to the Maryland Department of Natural Resources' *Maryland State*

Wildlife Action Plan for 2015–2025, rare and threatened species in the Jug Bay area include seven plants and eight animals. In addition, historical records exist for several rare plant species, including the state Extirpated thread-leaf naiad and long-stalked crowfoot, the state Endangered short's hedgehyssop, the state Rare / Watch Listed shoreline sedge, and the state Uncertain Status graybark grape. Unconfirmed records also exist for spongy lophotocarpus, a state Rare species. Surveys for these species have not been conducted recently, and a thorough effort may reveal that several still persist. The Baltimore checkerspot butterfly is now considered a state Rare species, though the pumpkin ash (one specimen is just off the boardwalk in the Sanctuary) has been down ranked from a state Rare / Watch Listed species to an apparently secure one.

The tidal hardwood swamp community lines the edge of the tidal marsh and is considered to be globally vulnerable to extinction, due to the restricted ranges of such swamps. These communities are found in tidal rivers in Virginia, Maryland, and Delaware, and less than a hundred occurrences of this habitat type have been documented worldwide.

12–15 *October* | CLIMATE AND LEAF COLOR

Jug Bay and its neighborhood have been sweating in 80°F weather and oppressive humidity, so it's a welcome change to enjoy cooler days, even if accompanied by heavy rain. It's a time when Marylanders take to the roads to see autumn in its best colors. Using the online Foliage Network, many head west to the Shenandoah Mountains or north to New England.

My wife and I have driven to a conference in Toronto, Ontario, so we have had an opportunity to observe autumn's effect on color as our journey progressed. Because northern areas had experienced prolonged high temperatures, lower New York State and the Toronto area have only had a "moderate" leaf drop, with color more likely to be on the trees than on the ground. The much needed rain of these past few days, however, may help the colors in leaves that are yet to turn in the Sanctuary. Meanwhile, the mountainous areas of upper New York State, Vermont, New Hampshire, and Maine are now in peak colors.

Drought, late season warmth, and cloudier days have become more common with climate change and have muted leaf colors in autumn. According to Barrett Rock of the University of New Hampshire, they now progress "from green to pale yellow, followed by leaf fall. Stressed trees have difficulty making antifungal compounds, which makes it hard for them to create beautiful bright colors."

Rock also observes that there was a time when Columbus Day meant visits to New England for spectacular colors. Now foliage changes may not have even begun by early October, and frost comes as late as early November.

The damp and dreary conditions in which we drove north gave way once again to summer-like temperatures on our return journey a few days later. Brisk northeasterly breezes have eased and light southeasterly to southwesterly winds have prevailed, bringing temperatures back to the upper 70s. They will decrease quickly as northerlies are predicted to sweep in a day or so later.

17 *October* | HUMMERS AND MORE

I watch three ruby-throated hummingbirds plundering a nearby flowerbed. They fly in all directions, forward, back, up, or down, and, when not chattering their chase notes, produce squeaks and twitters of an amusing kind. These are either young birds or females embarking on their long migration south; the males left in early August. Those seen one day are gone the next, to be replaced by new ones. It is said that the sources of nectar are only refueling stops, enabling the hummingbirds to search for insects and pile on fat. During migration, their normal body weight increases by over 50 percent, and they become distinctly pudgy. A study by Theodore Zenzal and Frank Moore has shown that only an additional 0.07 ounces of fat provides sufficient fuel for them to cross of the Gulf of Mexico, a journey of some 18–22 hours. Southern Mexico or Central America is their final destination, where they live a reclusive winter existence, competing with large numbers of flower feeders. We shall not see them again until next May.

Following a heavy rain, more leaves are hitting the ground, some moving like toy boats when reaching water. Tuliptrees continue to turn bright yellow, sassafrases are increasingly wine red, while American sweetgums present a kaleidoscope of bright red, orange, yellow, and purple hues. Only the wreath goldenrods and lingering great ragweeds challenge the decayed look of the flowerbeds. Fewer butterflies and bees now visit the beaten floribunda, and mosquitoes have become less annoying. At night I hear only a few plops of southern leopard frogs diving into the marsh, and bullfrogs utter a very occasional croak. The calls of crickets are fast becoming a desultory affair. Change beckons everywhere.

There's a marked decline in birds visiting the feeders outside the Wetlands Center, and those that do come as flocks, devouring the seeds voraciously. House finches, in their messy molt, have taken on a mean look, while the vanishing brilliance of male American goldfinches reminds me of the tatty appearance of caged birds.

18 October | BEES

A paper in *Nature* by Edwige Moyroud and colleagues shows that flowers can create a halo of blue that may be attractive to pollinating bees. As it's hard for flowers to manufacture that color, they use a trick of physics: tiny ridges on the petals alter the sunlight, creating the illusion of blue.

I must say that bees produce one of the sweetest sounds in all of nature.

AUTUMN MAGIC

On autumn days, when the nights are drawing in, I like to enjoy the last hour or so of daylight walking through the woods and along the marsh. The damp air is sweet, and nature's smells are more intense as I carefully wend my way along trails, hoping for a passing red fox, North American beaver, or muskrat. The dampness and decay of autumn have produced an abundance of mosses and liverworts that, along with beautiful and much smaller lichens and fungi, make striking, subtle mosaics on the woodland floor.

Clumps of pale stems push up through the rotting leaves, topped by waxy, inch-long flowers bearing four to five small white petals, facing earthward. They reminded colonists of the ceremonial pipes, or calumets, of New England's indigenous inhabitants, hence their name, indianpipe. I am lucky to see them, as they normally die back in the middle of fall. Once the flowers are pollinated and begin to form seeds, they will turn up to face the sun. Though lacking chlorophyll, indianpipes are a true flowering plant, relying on energy drawn from fungi in the woodland floor.

The marsh now has a yellow-brown tone, with water slowly reclaiming its wintery aspect. Muskrat lodges and osprey platforms stand out. Canada geese glide in and land midriver, and a few gulls wend their way to their roosts, their whiteness made golden by the setting sun.

I find myself drawn in by all this—the stillness, the slowly muting colors, and beauty the eye can hold and celebrate. My reverie is short lived as the foliage comes alive with the movements and soft rattling calls of Carolina chickadees. Other birds stir: American robins with their roosting alarm calls, Carolina wrens scolding from the undergrowth, and throaty "checks" from red-winged blackbirds finding roosts in the cattails. But it is the call of a frog, a spring peeper, that closes the day.

On my way back in the darkness, I catch the stirrings of an invisible barred owl and a fox. As the air is cool and still, it's possible to hear the faint calls of birds as they migrate in the darkness, high above. I recognize the cries of warblers, sparrows, and tanagers all headed for their wintering grounds in the tropics.

The calls of migrants are now being recorded, to estimate the numbers passing in the night sky. Ever since British ornithologist David Lack first picked up migrating birds when conducting radar research in 1941 (water in their bodies reflects radar beams), weather maps have been used routinely to shadow birds on the move. To identify individual species, Kyle Horton and others at the Cornell Lab of Ornithology are combining radar data with acoustics, thermal imaging, and citizen-science records (on eBird) to analyze such "avian clouds." Sightings logged by birders provide clues about which species might be passing through an area, thereby enabling the data to match patterns on weather maps. This work has begun to illuminate the composition, timing, density, direction, speed, and altitude of nocturnal movements of migratory birds.

The authors plan to use algorithms to define new flyways, as the current flyways are based upon waterfowl migrations. Their hope is that birders will be able to look at maps showing which species arrived overnight. Meantime, they can already trace continent-scale migrations and frequent flyers on any given date.

19 October | AQUATIC ECOLOGY: CREEK, MARSH, AND RIVER

> Instead of giving students scientific information, teachers need to help them develop skills that will make them scientifically literate adults.
>
> Daniel Levin, personal comment

Each year school groups visit Jug Bay, learning about the impacts of tides from dedicated research workers. As logistics make it difficult for students to make a followup visit, they only have a limited time to observe the natural environment and view organisms under changing conditions.

So it's fortunate that the predictability of tides at Jug Bay affords unique opportunities for the visiting students to investigate changes during a single day. Many factors affect the levels of nutrients and dissolved gases in the marsh, but the rise and fall of the tides can produce great changes, even over the course of a day.

The groups analyze different factors during low or high tides at a tidal location (the marsh) and a nontidal location (Two Run Branch). They also observe the plants and animals and make inferences about adaptions needed to survive under different conditions.

The students have found that the creek is unaffected by the tides, so the water level does not change. In contrast, the water level in the marsh varies about two feet between high and low tides. During low tide, the level of dissolved oxygen usually drops, especially during a heat wave, when oxygen-depleting nutrient levels in the water might be high. This can lead to the periodic fish kills that take place around the bay.

The differences in dissolved oxygen are largely caused by variations in temperature: when low tide occurs in the middle of the day and the sun is shining down on the marsh, the water temperature rises and its oxygen level decreases. The creek, which runs through woodland, has cooler water and, thus, a greater holding capacity.

When the students analyze the percentage of oxygen saturation that could be held at the respective temperatures at the two locales, they usually find the creek water at 90–95 percent, while the marsh water is generally only at 65 percent.[4]

Clearly, some factor other than temperature is contributing to the smaller amount of dissolved oxygen in the marsh. Low tide proves to be a good candidate: water is stagnant and fresh oxygen is not replenished, as it is during high tide. As an indicator of oxygen decline, mummichogs, that normally tolerate low levels of oxygen, can often be seen sucking air at the surface during low tide, trying to restore their depleted oxygen.

A subsequent study group at Otter Point has found that many factors, including sunlight, photosynthesis, wind, water velocity, pollution, tide, water depth, water temperature, and season, affect the amount of dissolved oxygen in the water. Oxygen saturation typically exceeds 100 percent between February and May and dips in late summer (being lowest at midday and highest before dawn). In early fall it is down to 70 percent, when water temperatures are higher and the maximum decomposition of plant matter occurs.

One continuing study by Jug Bay volunteers has monitored water clarity in the Patuxent River, finding a slight increase in clarity in the main channel over a 15-year period (since 1988). If this trend continues over the next decades, it could bring a significant benefit to the river and the marsh.

Sarah Greene performed nitrate analyses on water samples collected at high and low tide in the marsh and discovered that the levels at high tide exceeded those at low tide. This is largely due to nitrates carried to the marsh by the Patuxent River. It was also determined that in the spring and summer, the uptake of nitrates by phytoplankton and marsh plants, as well as denitrification (conversion of nitrate to atmospheric nitrogen) decreases nitrate concentrations in the marsh.

White perch are schooling in the Patuxent River and tributaries of Jug Bay, poised to leave for their main Chesapeake Bay wintering grounds.

22 *October* | SAPWELLS

I take the trail around Otter Pond and am asked by a visitor if I know the cause of "those horizontal holes in the bark of some trees." We walk to one of the trees and observe lattice-like patterns of one-quarter-inch shallow holes across its trunk. The culprits are yellow-bellied sapsuckers, pecking the holes during their winter sojourn at Jug Bay.

Using their brush-tipped tongues, sapsuckers sup the fluid oozing from the inner bark (phloem) of each hole, or sapwell. The sap is rich with nutrients being carried from the leaf tops, and each hole is diligently cleansed.[5] When the renewed flow later subsides, another row of holes are pecked above the existing row, with the inner cambium layer also being consumed. In addition, the sap provides nourishment for a host of insects, especially ants, which may constitute as much as a third of the sapsuckers' diet.

Sapsuckers have been found drilling sapwells in more than 250 species of trees and woody plants, though they have a strong preference for birches, maples, and hemlocks. In spring and summer, they favor young forests and edge habitat and will search for the best fluid flow. Fungi and bacteria sometimes invade the sap and can cause serious damage to the affected tree.

Hummingbirds also feed off the sapwells. Indeed, in some regions, ruby-throated hummingbirds rely so much on sapwells that they time their spring migration with the arrival of sapsuckers. Other birds, as well as bats and porcupines, also visit sapwells.

Yellow-bellied sapsuckers are the only woodpeckers in eastern North America that are completely migratory. Besides drilling sapwells, they also glean insects from tree trunks in more typical woodpecker fashion, as well as sally forth to catch insects in the air. Berries and fruits are eaten in all seasons.

24 *October* | LEAF SHAPE

As they flutter down, leaves have a story to tell about the climate each prefers and why those of separate species come in different shapes. A mathematical model by Ben Blonder and his colleagues now offers an intriguing answer, from the perspective of leaf veins. Recognizing that three factors serve a leaf—the amount of carbon dioxide needed to grow it, its lifespan, and its rate of photosynthesis—their model predicts how these factors best serve a plant, using three properties in the vein networks: density; the distance between veins; and the number of regions containing smaller veins, or loops, resembling capillaries in humans.

Vein density is a measure of how well the veins keep a leaf supplied with water and nutrients. The number of loops reflects the resilience of a leaf and is related to how long it lives, since they provide ways to reroute supplies should the leaf be damaged. The research workers have shown that leaves with a higher vein density (i.e., more plumbing) favor wetter environments, while also seeking out the brightest and warmest places. If a plant constantly requires large amounts of water, it could favor certain geometrical arrangements of veins and overall leaf shapes. So it's the veins—the skeleton of the leaf—that determine a classic maple shape or that of a blade-like willow. The large leaves and high vein density of basswoods intercept sunlight in the canopy very efficiently. They can support high rates of photosynthesis, because of their ability to tolerate high rates of transpiration. By contrast, yew leaves have just one single vein down the middle, resulting in a slow water supply and slow growth. These leaves probably don't get much wider or longer, because they are constrained by having only a single vein. They are adapted to low light and slow growth conditions, where this limitation is not an issue and, instead, allows them to outcompete other species.

From redwood trees looming out of the fog, to windswept hawthorns, to alders growing alongside a stream, each leaf has its own story to tell.[6]

25 *October* | CHIPMUNKS PREPARE FOR WINTER

The Latin name *Tamias striatus*—"treasurer, or storer" and "striped, or furrowed"—is a very apt designation for eastern chipmunks. I see them as they scurry across the woodland floor, preparing for winter with bulging cheek pouches. Their dark brown side stripes contrast with light brown stripes along the back and their black tails. An acute sense of smell enables them to find nuts, including those of American beeches, hickories, hazel alders, black cherries, and oaks. Michael Caduto determined that the pouches have a maximum capacity of 70 sunflower seeds, 31 corn kernels, or 12 acorns, the food being pushed through gaps between the lateral teeth. The aim is to store enough supplies in extensive and deep underground tunnels, places where chipmunks will enter into long periods of torpor during the winter. By the end of autumn, this larder may contain 5,000–6,000 nuts. Chipmunks, however, are omnivorous and will consume berries, fruits, and mushrooms. On occasion, they will eat invertebrates and even small mammals. Scattered hoards may offer an alternative food supply if the chipmunk's main food source is stolen or destroyed. All stores provide them with crucial nourishment during the winter and early spring and, occasionally, when nut and seed crops fail.

Curled up in a bed of leaves in winter, a chipmunk will stir occasionally to raid its larder and, on mild days, venture aboveground. As spring advances, chipmunks will emerge from their burrows to mate. Some four to five weeks later, four or five young are born and in a further five weeks will become independent.

Chipmunks prefer beechnuts, but the bark of the beech tree is smooth and makes it hard for them to climb, especially when the crop is poor.[7] To overcome this impediment, neighboring trees provide access to the beech canopy and the clusters of beechnuts.

26 *October* | BIRD COUNT AND ECOLOGICAL ENGINEERING

A cold day, and the bird count group has come appropriately attired as it surveys the marsh and the Patuxent River from the Observation Deck. Dying brown leaves and cattail stalks with fluffy tops sway in the northwesterly breeze,

The Wider Benefits of Caching

The development of massive ice sheets and their movement south during the last Ice Age some 10,000 years ago displaced many species, including trees. For decades, biogeographers wrestled with understanding how tree species migrated north and re-formed the enormous Austroriparian forest of eastern North America so rapidly when the glaciers melted and retreated. Historical information indicates that the oak forests moved northward at an average pace of 380 yards per year. By transporting nuts up to several miles from host trees, blue jays may well have responsibility for this spreading germination.

Source: Johnson and Adkisson 1986.

as do muddy spatterdocks poking above the high tide that gently rocks stands of fallen wild rice. On the shore, red maple trees adorned in scarlet advertise their preparation for winter.

Along the river and the marsh edges, gulls effortlessly ride the wind, repeatedly circling over the water and back toward the cattails. Sometimes the gulls concede slowly to air currents, and at other times quickly, floating up and down in an astonishing balancing act. On occasion, they rise high enough for their radiant white bodies to catch the piercing blue sky.

Waterfowl have arrived these past several days. We count 500 Canada geese, sizable flocks of mallards, and 85 American black ducks. There is also a delightful little pied-billed grebe exploring the marsh's edge. A merlin whizzes by in one direction, a Wilson's snipe in another. Bald eagles assert their magnificence against the sky, but it's the passage of northern harriers, a sharp-shinned hawk, and a red-shouldered hawk that affirms the deepening of migration.

Later, along the Railroad Bed Trail, the autumn push is confirmed by the strong presence of eastern towhees and various sparrows: white-throated, song, and swamp. Red-winged blackbirds congregate, and in their midst are rusty blackbirds. Soras call from the cattail stands. We've heard reports of hundreds of yellow-rumped warblers in the region and are pleased to find them in sizable numbers in the trees bordering the trail.

On the return journey, I am struck by the musical crunching beneath our footwear. Rummaging among the leaves, I find the acorns of white, red, and scarlet oaks and the nuts of mockernut hickory and American sweetgum. As I do this, some blue jays squawk, as if to remind me that they are still selecting and caching some of these acorns for the coming winter, when food supplies are expected to run low. Their efforts are truly prodigious and have the benefit of growing new oak trees and regenerating forests when the buried acorns are not retrieved.[8]

The jay is not alone in stashing food. Chickadees do the same. Studies on the growth of the hippocampal region of a bird's brain in late summer and early fall have shown its importance in spatial memory, namely where to find the seeds. A *New York Times* article by Natalie Angier describes how Fernando Nottebohm and his coworkers focused their research on the remarkable ability of black-capped chickadees to recall the locations of hundreds of stored seeds. Their findings produced the first evidence that some neurons in the chickadees' adult brains are replaced periodically, accompanied by the birds learning new behaviors.

The scientists also suggest that as demand for memory space peaks, chickadees discard cells that hold old memories and replace them with new cells that store fresh ones. They and others have noted that studying the ability of a bird's brain to generate new neurons might uncover ways to replace brain cells that were lost due to injury, a stroke, or degeneration, as happens in diseases such as Parkinson's, Huntington's, and Alzheimer's.

29 *October* | A LIFE WELL SPENT: JOHN TAYLOR, ARTIST AND CONSERVATIONIST, DIES

In one painting, an osprey stands erect on a dead tree limb, a fish held firm in its impressive claws, while a mate glides nearby. Another portrays a great blue heron, also perched erect on a dead limb. Both depictions, hanging in the Wetlands Center, allow the observer to admire the deft portraits of the birds and the evocative setting, the Jug Bay marsh.

The artist, John "Bud" Taylor, who died on this day, spent his life in the marshes and waterways—and with the wildfowl—of Chesapeake Bay, capturing landscapes and wildlife. His fascination with birds and art began in the fourth grade, when a teacher formed a Junior Audubon Club. From then on, he drew and painted birds, but the idea of becoming a professional artist did not develop until he began work for the Division of Birds at the Smithsonian Institution's National Museum. Here, John developed a close association with artists and ornithologists that provided the basis and inspiration for his life's work.

Following military service, during which time he visited and studied with bird artists in England and Germany, he returned to the Smithsonian and studied art and design at the Corcoran School of Art. He then accepted a position as artist and editor of the State of Maryland's wildlife publication, the *Maryland Conservationist*.

John was recognized for his artistry through commissions from the National Geographic Society and the National Wildlife Federation. He designed Maryland's first deer and trout stamps and was selected as the artist for the first state waterfowl stamp in 1974. He also won the Maryland duck stamp contest in 1979 and 1984. His life and art were documented on public television in a segment of *Maryland Outdoors*.

Taylor's *Birds of the Chesapeake Bay* includes 40 reproduced paintings in full color, as well as text adapted from his field diaries. A second book, *Chesapeake Spring*, also based on his field diaries, features 68 works in color, depicting all forms of life on Chesapeake Bay.

During his retirement, John worked mostly in a studio overlooking beautiful Pennington Pond, an inlet just off Chesapeake Bay, where he enjoyed the presence of his seasonal companions.

John donated his field diaries and much of his book collection to the Smithsonian. Jug Bay was his "little corner of the planet," and, as Christopher Swarth (the Sanctuary's director from 1989 to 2012) has observed, John's field sketches and diaries provide a "valuable record of a changing landscape and its dwindling bird populations."

In his paintings on display at the Wetlands Center, John captured the unique qualities of both herons and ospreys, along with the Jug Bay landscape. Many of us will now look at them anew and recall this remarkable artist and dedicated conservationist.

October says goodbye to summer's short lease of balmy days. It has clung to a fading beauty as lingering flowers succumb to colder nights. Migrants have left for warmer climes, to be replaced by waterfowl and other birds content to face winter's conditions. The month is a pause between the certainties of summer and winter, allowing wildlife to deepen their preparations for winter.

A woodland slope showing leaf fall and beech trees still leaved. *Jug Bay Wetlands Sanctuary*

NOVEMBER

The Beaver Moon

The thinnest yellow light of November is more warming and exhilarating than any wine they tell of. The mite which November contributes becomes equal in value to the bounty of July.

Henry David Thoreau, *Journals*, 25 November 1857

OH, THE CLIMATE

A warm dome of high pressure at high altitudes resided over the eastern United States in October, with the consequence that the area posted its third-warmest October on record—5.5 degrees hotter than normal, with only last-minute cooling preventing it from becoming the warmest on record. All but the months of May and August have been toastier than normal, and February and April ranked as the warmest on record. So far, the overall temperatures this year rank second only to those in 2012.

Not only was it unusually warm, but also quite dry. The monthly precipitation total of 2.02 inches was 1.38 inches below normal, and it ranked as the sixth driest of the 2000s. Only the coastal storm at the month's end, which dispensed 1.06 inches during 29–30 October, saved it from being even drier.

The combination of weather conditions in September, when rainfall was 2.29 inches below normal, and October has marginally pushed the year's precipitation total into below normal territory.

1 November | PEEPING AND SLEEPING

Spring peepers never cease to amaze. After their loud acclaim during early spring, they busy themselves with breeding, and in the following months are only heard from time to time. Now, with unusually warm conditions (74°F), the calls of some males return as they prepare for winter and the next breeding season.

How do these tiny, one-inch marvels survive the challenges of winter, especially when freezing conditions prevail? Though they hibernate under logs or rocks or behind loose bark on trees, such hibernacula (winter refuges) are insufficiently protected from cold weather, so the frogs may freeze, along with other inhabitants of such shelters. To endure, water exits the peepers' cells, and the remaining glucose fluid acts as a natural antifreeze to help preserve vital organs. Up to 70 percent of the frog's body can freeze, to the point that its breathing and heartbeat stops, and it appears to be dead.

Scientists still aren't sure how frozen frogs wake up, but once they thaw out, most frogs will go through a period of healing before resuming their normal lives. To enhance their survival, these freeze-resistant peepers emerge earlier in the spring than their predators.

Current evidence suggests that freezing episodes in five species of North American frogs are frequent and, in some cases, may be prolonged. Scientists Jack Layne and Richard Lee found that repeated freeze/thaw episodes deplete the frogs' glucose reserves, owing to a persistent loss via renal excretion and greater uptake to rectify metabolic disturbances caused during freezing. Nonetheless, these astonishing creatures are often abundant in their respective natural communities. An alternative strategy of selecting hibernacula to avoid subzero temperatures also has energy costs for the frogs, incurred both by burrowing to considerable depths and by prolonged exposure to higher environmental temperatures.

2 November | EARLY MORNING AND THE MIGRATION OF MARBLED SALAMANDERS

The stars dissolve in the pale pink of dawn, slowly losing their reflection in the gently moving river. A flock of Canada geese breaks the cold night's silence with their splattering takeoffs and noisy honking. Near and far, fish crows chorus, and the watery cries of American coots cross the open water.

Light suddenly floods in as the sun rises, bronzing the standing cattails and caressing the wondrous palette of the distant trees. The white rump of a northern harrier stands out as the bird flies low over the marsh in languid and gliding pulses.

In the marsh, the muddy, folded leaves of spatterdocks are all that remain of the live vegetation. A common winterberry bush, curiously mixing with the scarlet-and-yellow-leaved highbush blueberries, bends toward the Railroad Bed Trail,

A marbled salamander. *Kerry Wixted*

heavily laden with red berries. Scat at the top of an otter path crossing the trail is white, with crayfish remains revealing the otters' seasonal switch from a diet of fish. The beaver lodge near the Observation Blind is now an impressive mound of large branches, held together by plastered mud.

In open spaces of the upland forest, beechnuts are raining down on the leafy floor, now crunchy to the tread. The trees' lower branches still hang on to the husks enclosing the nuts. Eastern gray squirrels, voles, and birds will doubtless benefit from such plenitude.

Much as we marvel at this fall's migration of birds, another is unfolding. Marbled salamanders have been emerging from their subterranean habitats in the upland woods, moving to pools nourished by late seasonal rains, in order to begin their mating and annual breeding cycle. Studies conducted at Jug Bay by Karyn Molines confirm that an increased amount of rain results in a greater number of salamanders on the move. Their peak shifting normally occurs in mid-September, trailing off in late October to early November. This year's dry months have delayed their migration, however, and I join Chuck Hatcher, a Sanctuary volunteer, to see if we can find nests of breeding females.

We edge around the slopes of a vernal pool newly replenished by last weekend's downpour. Chuck turns over logs very carefully and eventually uncovers a nest full of 100 or so eggs. They are a dull black, peppercorn sized, and are neatly packed in a shallow depression in the soil and leaf litter. Under another log, we find a female attending her clutch. She remains still and allows us to admire the silvery gray bands across her lustrous blue-black body.

David Scott's study of the trapping histories at a wet forest site over a 12-year period reveals intriguing patterns of salamander movements. To court and mate, on rainy nights in September and October, both sexes tend to begin significant shifts to their breeding areas around vernal pools, even though the ponds themselves may not yet be filled with water. After laying her eggs, the female stays with her developing clutch until rain fills the pool and triggers hatching. By late November, the salamanders leave the vernal pools and return to the upland forest, their annual breeding migration coming to an end.[1]

The length of incubation will vary. Eggs usually hatch in autumn, but if rain never comes, the eggs can overwinter until spring to hatch. The aquatic larvae take from two to nine months to metamorphose into terrestrial juveniles. These juveniles rest under logs, stones, rocks, and debris in fairly shaded situations, and they later hibernate in deep subterranean burrows. About another 15 months are needed for the juveniles to reach maturity.

Though rainfall is necessary for vernal pools, seasonality plays a strong role in determining the magnitude of the migration. Scott has hypothesized that the timing of migration is predictable from year to year, because marbled salamanders are not dependent on water to lay their eggs. The mass migration of most salamanders during the mid-September window may indicate other environmental, behavioral, or physiological adaptations not measurable through the trapping study. Once mating is completed, the salamanders are more casual in returning to their nonbreeding habitats, resulting in a more prolonged and less intensive number of salamanders after each rainfall event.

4 November | BALTIMORE'S HARRIET OSPREY

The departure of ospreys from Jug Bay invites speculation about their migration to winter quarters. Tracking devices now provide clues, and Harriet, an osprey that nested in Baltimore, has given us our best insight to date. She left Baltimore on 7 September, and by 10 September was recorded along the coast of

Georgia, about 50 miles south of Savannah and 150 miles northeast of Hurricane Irma. The multiple satellite tracker readings along the Georgia coastline indicated that she waited out the higher winds generated from Irma, farther south, before resuming her flight.

By 16 September, Harriet had flown approximately 1,126 miles in 10 days, with the tracker locating her in Boot Key, Florida. With Hurricanes Jose and Maria potentially in her path, she again waited out the bad weather. On 23 September, Harriet reached Cuba; on 28 September, Haiti; and the Dominican Republic on 3 October. Between 11 October and 30 October, the transmitter indicated that Harriet was in Puerto Rico, having flown 2,200 miles.

The next leg of her migration would be the most challenging—crossing some 500 miles of open water to en route to her South American wintering ground. Alas, no satellite transmission was received after 4 November (near San Lorenzo, Puerto Rico). Possibly Harriet encountered bad weather, perhaps the solar-powered battery did not receive sufficient energy for the transmitter to send out a perceptible signal, or she might have been the victim of some unfortunate circumstance and perished. Note: The last scenario is considered most likely, as she did not return to the same area to nest the following year (2018).

5 November | MARCHING ON

Events have moved quickly these past few days. No longer do we enjoy the evening calls of insects. A chill infuses the air, and fog gathers over the river and the marsh. As evening closes in, a flock of slate-colored dark-eyed juncos, newly arrived from the north, explore the forest floor, and resonant honking flights of Canada geese affirm the relentless changing of the seasons.

8 November | LIFE IN RUNNING WATER

After the steady rainfall of the last few days, I find myself standing on the banks of the winding Two Run and Pindell Branches, mesmerized by the gurgling and churning water as it heads toward the Patuxent River. I was in the same area a few days ago, and much has changed. Where there were small trickles, there are now eddies (circular, swirling, vortices created as water passes obstacles, which Leonardo da Vinci likened to the flowing curls of a woman's braided hair), riffles (shallow water, in steep areas, breaking into foaming whitewater), pillows (upstream bulges, created whenever a strong flow meets a rock or other obstacle), and finally, pools (deeper, slower areas). Smooth glides of water are evident between the pools and riffles.

Within this galaxy of physical environments, there are microhabitats for many plants and animals. In rushing water, stones are overgrown with algae and mosses and support a rich assemblage of animal life. In contrast, the stones in slower-moving water have a relative paucity of both plant and animal populations. The favorable effects of stronger currents provide an environment physiologically richer in oxygen and nutrients. A slower current, however, may induce organic material to settle on the river bottom, and abundant animal life may emerge from these deposits.

The adaptations of life forms to the current are striking, for they have to resist its mechanical forces. Some plants have developed powerful holdfast organs; others, a flattened aspect; or still others, a reliance on gelatinous or calcareous tubes. Animals adopt through extensive compression, with some mayflies evolving to have fish-shaped bodies. Caddisflies flatten their larval cases and make them heavier by attaching little wings. Dipteran (two-winged insect) larvae and some snails possess powerful suction cups. Finally, there are the cavity dwellers, wholly dependent on contact to provide their stimuli.

As an academic, I had the good fortune to conduct research on fast- and slow-moving aquatic ecosystems and bring students into a new world. Now, as I leave the banks of the creeks, I recall those times with immense pleasure and take pride in the fact that some of these students became aquatic biologists.

Rain has returned, and leaves are bouncing back as gravity prevails and their tips lose water.

10–11 November | First Frost

The frost and freezing conditions of the past two mornings have altered the appearance of many plants. Some have become limp, blackened, and distorted, seemingly made worse where plants face the morning sun, thus defrosting quickly and rupturing their cell walls. Strong cold winds remove moisture from evergreen foliage more quickly than it can be replenished, and repeated freezing and thawing, or very rapid thawing, can be especially destructive.

The formation of ice within a plant appears to be the most damaging effect, not the sudden change in temperature. A cell's liquid contents expand as they freeze and can rupture the cell wall. If temperatures fall more slowly, the liquid in the cell has a longer amount of time to be forced outside through the cell wall, pushing the plant cells apart and dehydrating the cells. If the liquid then freezes, the resultant damage is called extracellular. Even plants that tolerate frost, however, may be damaged once temperatures drop even lower (below 25°F).

Most hardy perennials become dormant after the first frosts. The entire visible portion of the plant may turn completely brown until spring's warmth, or the plant may drop all of its leaves and flowers, leaving only the stem and stalk. Evergreens, such as pine trees, withstand frost, although all or most growth stops. A combination of low temperatures and heat from the winter sun results in frost cracks in tree bark.[2]

11 November | WINTER WHIPLASH

A few days ago the clocks were set back, and yesterday the mercury registered winter conditions—some 3.4 degrees above normal for November—ending autumn's balmy run. By 9:00 pm, the thermometer reading slid to 35°F, the coldest temperature in eight months, and more cold is forecast.

12 November | WITCHHAZEL, OR SNAPPING HAZEL: AUTUMN'S METRONOME AND ARTILLERY?

Any lingering flowers are now succumbing to nightly frosts, save for tall American witchhazels in the wet understory. Their spidery yellow petals, which I am now carefully viewing, may well survive until late November and provide nectar for still-active autumn insects. I use the word "carefully," because while the blossoms add some visual relief to the otherwise dying landscape, witchhazels are unique in another respect. They bear buds, flowers, and fruit simultaneously, the latter packing a punch.

In October, witchhazel fruits are ripe, containing two long, glossy black seeds in the pods of last year's flowers.[3] On reaching maturity during frosts, about eight months after flowering, each pod splits open explosively (dehisces), ejecting the seeds with sufficient force to fly up to 30–40 feet. As Thoreau wrote in his *Journals* on 23 September 1859, "I heard in the night a snapping sound, and the fall of some small body on the floor from time to time. In the morning I found it was produced by the witch-hazel nuts on my desk springing open and casting their seeds quite across my chamber, hard and strong as these nuts were."

Other examples at Jug Bay include jewelweeds, their other common name being spotted-touch-me-nots, referring to the dehiscence of the seed capsules, sending their contents many feet away. The expelling force, however, depends upon the type of flower: seeds from open flowers are tossed farther from the parent plant than seeds from closed flowers. Short-tossed seeds have genes from a single parent (a plant with both male and female parts) and land in spots similar to the parental habitat, whereas farther-tossed seeds, carrying genes from two parents, are better equipped to survive in conditions differing somewhat from the parental site.

Sphagnum mosses (peat moss) disperse the spores by wind. As their spore capsules dry, compressed air forces off the lids with an audible "pop," releasing the spores at up to 11 feet per second, in what is, not surprisingly, termed an air-gun mechanism. High-speed photography has shown that rings of spinning gases are created during the discharge, enabling the spores to reach a height of four to eight inches, farther than would be expected by ballistics alone. The acceleration of the spores has been measured at 36,000G. Such attributes may explain the wide distribution of sphagnum moss species.

All such efforts pale, however, compared with the sandbox tree, which can fling seeds 150–330 feet. An alternative name for this species is dynamite tree, due to the loud sound of its explosive dehiscence.

14 *November* | WINTER LOOMS

A cold northeasterly wind rattles the dying leaves of beech and swamp chestnut oak trees, buffets cattails, and ripples the surface waters of the river and marsh, which are increasingly indistinguishable. The yellowing stalks and decaying leaves of spatterdocks lie limp and gently rocking on the outgoing tidal waters. A lone double-crested cormorant heads downriver, and a flock of red-winged blackbirds eke out any remaining wild rice seeds.

The woodland is silent, save for the drumming of a northern flicker, chattering Carolina chickadees, and the gentle "chink" of white-throated sparrows. Downriver from Lower Glebe Marsh, a mixed flock of gulls circles endlessly, with their raucous cries echoing across the open waters.

Nothing breaks the stillness of the surface of the Beaver Pond, green with common duckweeds. In summertime, this blanket of vegetation is broken by the poking heads of turtles or gliding North American beavers. Now these flat, floating, colony-forming bodies, with no true stems or leaves, are producing winter buds (turions) that will sink to the bottom and remain there until spring. Then, each bud expels a gas bubble, causing it to rise to the surface, where it rapidly develops a new thallus, or frond. This high-protein plant is readily consumed by waterfowl and other animals, though the way it covers the water's surface threatens to shade the water column and impair photosynthesis in other plants.

Duckweed mats host a variety of small insects—feeding, laying eggs, or merely sheltering—that provide food for fish and frogs. Beavers in the pond will carry the fronds (thalli) in their dense fur, and waterfowl on their feet and feathers, transporting this plant to other freshwater bodies.

American hollies around the edges of the pond stand out from the now leafless trees and help relieve the area's increasingly monochrome tone. A chattering dense flock of common grackles, swooping in unison over the grasslands of the River Farm and landing in trees, now absorbs my attention. The birds create incredible fluid shapes of freewheeling, pulsating curves. It is a murmeration, a term associated more with a flock of European starlings. They create what's been described as "suspended reality" or, alternatively, as equations of "critical transitions"—systems poised to tip and be almost instantly and completely transformed. Each bird is connected to every other, so when a flock turns in unison, it's deemed a phase transition.

The physiological mechanisms allowing turning to be nearly simultaneous in birds separated by hundreds of feet and in the midst of hundreds of other birds remain to be discovered. European starlings may simply be the most visible and beautiful example of a biological phenomenon that seems to operate in proteins and neurons, hinting at universal principles not yet understood.

A great blue heron ponderously flying down the creek abruptly ends my delight, and I head to the Observation Blind to catch the sunset. Here, I hear North American river otters scurrying away in the cattails and observe one sinuously swimming and then diving to enter its lodge, a few yards from where I stand. The lodge is also now host to a muskrat.

15 November | Winter Is Coming Later, and It's Leaving Ever Earlier

The Weather Underground website reports that as of the end of October, 25 percent fewer states had a freeze than in normal years. Scientists say it is yet another sign of the changing climate, noting that a shorter freezing season has good and bad consequences: there could be more fruits and vegetables, but also more allergies and pests, as well as cascading effects on plant and animal interactions.

According to data compiled by Ken Kunkel, a meteorologist at the National Oceanic and Atmospheric Administration's National Centers for Environmental Information, the average first freeze from 2007 to 2016 was a week later than the average from 1971 to 1980—before the trend became noticeable. Overall, the US freeze season in 2016 was more than a month shorter than that in 1916. It was most extreme in the Pacific Northwest. Oregon's freeze season was 61 days—two full months—shorter than normal.

Global climate change has helped push the first frosts later. Also at play are normal short-term changes in air circulation patterns—although they, too, may be influenced by man-made climate change, producing a certain amount of chill. In an article by Seth Borenstein, Theresa Crimmins, a University of Arizona ecologist, notes another disturbing trend: pests that attack trees and spread disease aren't being killed off as early as usual. Other changes in climate have also been observed. In New England, many trees aren't developing their vibrant fall colors as they normally do, or used to do, because some take their cues regarding when to turn from temperature changes. Clusters of late-emerging monarch butterflies are being found much farther north than normal for this time of year and are unlikely to survive their migration to Mexico. Shorter winters and hotter temperatures could also lead to rising sea levels, with the probability of worse flooding during heavy storms. In Borenstein's article, Boston University biology professor Richard Primack tersely sums it up, "The long-term consequences are really negative."

17 November | Return of Migrant Waterfowl

Canada geese heralded the return of migrant waterfowl from their northern breeding grounds a few weeks ago, and on their tails have come many other migrant waterbirds. From the Observation Deck, the bird count group has noted

163 American black ducks, 150 northern pintails, a green-winged teal, and two buffleheads. All are a splendid sight, having recently molted into their winter breeding plumage. More will come here, as well as to other tributaries of Chesapeake Bay, the bay itself, and off ocean beaches.

A little beyond the geese and ducks, five majestic tundra swans are spotted. We have not seen them since the middle of spring, and they hold us spellbound. Formerly known as whistling swans, they carry their long necks straighter than mute swans, and their high-pitched, quavering "wow-how-oo" contrasts with the latter's weak bugling. They have returned from the High Arctic to winter along the Atlantic Coast. Flocks of tundra swans journey at high altitudes and may cover vast distances, flying day and night before taking rest stops.

Two populations exist in North America: an eastern population (some 60%) that winters mainly in the Chesapeake Bay and coastal estuaries, breeding in the Northwest Territories and Hudson Bay; and a western population (about 40%) that winters in the San Francisco Bay area and inland sites, breeding mainly in Alaska and northeastern Canada.[4]

19 November | SHREWS: A TALE TO TELL

Shrews are among the smallest and most retiring of North American mammals. It's now at the very end of their normal breeding season, and I am therefore quite lucky to see a northern short-tailed shrew. Its mouse-sized body, with a pointed snout, wraps a bundle of energy in soft fur. It is known as a voracious insectivore and is one of the few venomous mammals. Because of its extremely limited vision, a shrew has to rely on a series of ultrasonic "clicks" to make its way around the tunnels and burrows it has dug.

In the dying light, I stand on fallen leaves and bushy plant cover at the boundary between woodland and marsh—one of this shrew's habitats. I look for its resting and breeding nests and find the latter underneath a decaying log. The nest is lined with leaves, grasses, and some fur (possibly from meadow voles, common to Jug Bay). It contains four well-advanced young, with sleek, slate-colored fur, pinkish paws, and stubby tails. I back away, on the presumption that the parent shrew is likely to have a leafy burrow system for food storage that could be damaged by my tread.

Unlike other shrews, the short-tails spend most of their time underground,

A northern short-tailed shrew. *Giles Gonthier*

rooting for food beneath the leaf litter on the forest floor. They prefer river flood-plains and sandy soils near swamps and marshes, only venturing into grassy areas during or after rain. They are mostly solitary, as well as pugnacious, exhibiting aggressive displays and vocalizations when encountering other members of their species. Daytime involves highly active periods, lasting about four to five minutes, followed by rest periods, averaging 25 minutes.

Northern short-tailed shrews consume up to three times their weight each day, preferring insects, earthworms, snails, and other invertebrates. On occasion, they will attack prey larger than themselves, including mice and even other shrews. Though active during cloudy days, they forage mostly for a few hours following sunset.

For the approaching winter, they often hoard food, with a study by Joseph Merritt finding that a northern short-tailed shrew caches 87 percent of the prey it seizes. They do not hibernate, and their ability to consume almost anything they catch allows them to survive cold winters. Other winter adaptations include constructing a lined nest to conserve heat, foraging below the leaf litter or snow (where the temperature is milder than on top of a snowy surface), and reducing their activity levels during cold periods. These shrews are also able to increase

On Venom

Northern short-tailed shrews have poisonous saliva, enabling them to kill mice and larger prey and paralyze invertebrates (such as snails), storing them alive for later eating. The venom is secreted from submaxillary glands in the jaw, at the base of the lower incisors, and is transmitted in saliva as shrews bite their prey. The saliva contains a kallikrein-like protease (an enzyme capable to cleaving bonds in proteins), used to paralyze and subdue prey. The toxin is strong enough to kill small animals up to sizes that are somewhat larger than the shrew itself. It can also result in painful bites to humans.

The role of this venom appears to be related to three aspects of the northern short-tailed shrew's life:

they have a high metabolic rate, which requires considerable amounts of food;
they hoard most prey, instead of immediately consuming it; and
they rely extensively on invertebrate food.

It is hypothesized that the venom is used to immobilize insects and other invertebrates, thereby providing a ready source of fresh food. Poisoned insects remain alive, but lethargic, for up to 10 days as a fresh source of food, supplying a cache of comatose insects with which the shrew can meet its future metabolic needs.

Sources: Kita et al. 2004; Martin 1981.

their ability to generate body heat by nonshivering thermogenesis in their brown fat tissue.

Despite these capabilities, evidence suggests that less than 10 percent survive the winter.

21 *November* | In Search of "Little Brown Jobs"

Many areas bordering the Patuxent River have been the source of gravel for the construction industry. When extraction ceases, most of the sites are abandoned and ignored. Some, such as Wooton's Landing (upriver, and part of the Jug Bay Wetlands Sanctuary complex), become opportunities for the State of Maryland to restore natural habitat and other ecological services. Here, following 15 years of such efforts, marshland and ponds have been created, interlaced with maturing woodland and patches of grassland. With so many habitat transitions, it is an ideal place for observing birds under a bright clear sky, and its sparrows are my targets today.

Most people find sparrows to be boring "little brown jobs," but during winter they, along with waterfowl, bring welcome relief to an often bleak landscape. They're far from simply brown, and their subtle, complex plumage patterns vary according to their habitat and maturity. Identifying field marks poses a major challenge, but it's their skulking nature that spells a particular difficulty. So it is with patience and close observation that birders turn to shape, behavior, and location for a diagnosis. In the breeding season, habitat choice is often a reliable guide; in winter, it's less exacting and often confounds observers.

I begin with a large, chunky bird, warily scratching away at dead leaves in woodland habitat. Its towhee-like kicking up of leaves in search of food in leaf mold and its rusty rump, back, and tail make it clearly a fox sparrow. This one is a red subspecies, found in the northern and eastern portions of North America, and is one of three or four fairly distinct groups; some taxonomists consider them to be separate species. A similar bird emerges in clumsy flight from weedy undergrowth, its long, rounded tail flipping to one side with a pumping action, continued on landing. It has an eye line and eyebrow, and the heavy streaking on its breast converges into a large central spot. Though other sparrows also have a spot, this one is unmistakably a song sparrow, with its moderately contrasting breast streaking.

I pass marshy terrain and, true to form, out pop a pair of swamp sparrows. Known to be ground feeders, they're usually found in low, open, wet areas at pond and marsh edges. They, too, bob their tails as they fly away, but they don't go far. They are plump little birds, with a rusty overall color, white throat, and buffy breast (matching the winter colors of cattails), and they keep close to the damp earth. Swamp sparrows arrive in the Jug Bay area during late September to early October and are common residents in the tidal wetland until their departure in April.

Next come the white-crowned sparrows, with their attractive head markings of alternating black-and-white stripes. The white-throated sparrows are immediately recognizable by their unmistakable song, "pure, sweet, Canada-Canada-Canada," coming from a woodland opening of low shrubs and small trees. Delivered in a minor key against a biting northwesterly wind, it warms the heart. They're squat little birds, with a gray bill, yellow eyebrow, and very white throat. In contrast, the lone white-crowned sparrow emerging from bushy cover adjacent to open grassy areas is tall and erect, almost regal, with a gray throat and pinkish bill. This sparrow is one of the most studied in North America, especially for its song dialects.

Along the many trails, dark-eyed juncos, with unmistakable white outer tail feathers, dive into thickets. These dapper gray and white birds with pink bills are the slate-colored subspecies, found only in eastern North America.

I'm growing cold and begin to trek back to the warmth of my car. Before reaching it, however, on the edge of the path beneath some bushes, I come across a small, slim bird with a thin long tail, rufous crown, dark eye line, and dirty grayish breast and rump. It's a chipping sparrow, on its way south. Before the house sparrow's invasion of North America, the "chippy" was regarded as the most domestic of all sparrows. Indeed, Audubon found it "almost as abundant in our country as the Domestic Sparrow in Europe" and "nearly as familiar, though otherwise different in habits."

In midmorning I move to natural grassland habitat near Sands Road, formerly a landfill area. Portions are mown in late August to halt tree growth and avoid disrupting breeding activities. The open expanse of the grassy meadow is classic savannah sparrow territory, but not until flushing one does it reveal its distinctive, undulating flight and small, short, square tail before quickly dropping down and disappearing into the grass. With dull brown plumage and a streaked breast, it's well concealed in the long grass. I disturb many more as I cross the meadow.

Before the morning is out, I spot one more sparrow near the grassland. It's a slim bird, with a rusty cap, pink bill, and unstreaked underparts, but the clincher is its white eye ring, giving it a wide-eyed look. It's a field sparrow. It flies to the top of a grass stalk, bending it sufficiently to the ground to feed on the seedhead. What a delightful bird, but how sad to hear that its population is declining in North America, because the intensification of agriculture and urban development are destroying its breeding habitat.

23 November | DEAD TREES NOT SO DEAD

 No dead tree is ever lifeless. Indeed, few other features in a woodland or forest are likely to support more animal life. In North America, about 80 species of

Snags for Wildlife

A snag habitat emerges when a large tree dies and forms what is known as a "hard snag." As this hard snag decays, it gradually becomes a "soft snag." Hard snags tend to have their bark intact, while the heartwood (the nonliving inner core) and sapwood (the younger, softer, growing wood between the bark and the heartwood) remain firm. Both kinds of snags are sought by cavity-nesting birds. A soft snag has considerable decay in its heart- and sapwood. As fungi infiltrate the heartwood, the tree becomes soft or hollow, and it eventually falls over.

Snags harbor many insects that prove food for wildlife. Birds such as brown creepers, red- and white-breasted nuthatches, and woodpeckers forage in the snag's outer surface for beetles, spiders, and ants. Woodpeckers eat the larvae and pupae of insects found in the inner bark, and mammals (such as raccoons and American black bears) may explore these areas to harvest protein-rich insects. Strong excavators, such as pileated woodpeckers, prey on carpenter ants and termites in the heartwood. The space between partially detached bark and the tree trunk is where red- and white-breasted nuthatches, wrens, and brown creepers roost or search for food. Pacific tree frogs, several species of bats, and many butterflies also shelter there. Mosses, fungi, and lichens can be found growing on logs and provide food and microhabitats for various fauna.

Sources: National Wildlife Federation [undated]; US Forest Service 1988.

birds, over 40 mammal species, and some dozen reptiles and amphibians rely on standing dead or dying trees for resting, shelter, roosting, denning, foraging, singing, courtship, mating, nesting, and other critical functions. In addition, dozens of invertebrates—worms, insects, spiders, and more—also regard snags as their habitat. When such trees occur along streams and shorelines, they may fall and add woody debris to an aquatic habitat. In all, says the US Forest Service, some 1,200 species of fauna rely on dead, dying, or rotted hollow trees.

Turtles, snakes, and other cold-blooded species often bask atop the logs of fallen trees. Though considered primarily to be a thermoregulatory matter, basking is now also thought to help turtles acquire vitamin D.

Fallen snags form habitats inside and under the places where they lie as logs or branches. They provide food and shelter for the many beetles and bees that dig tunnels in the decaying wood. Bark beetles live directly under the bark, leaving behind galleries, or tunnels, and colonies of termites and carpenter ants chew through the wood to build their chambers. Hollow logs afford cover, and sometimes dens, for small mammals such as foxes, rabbits, skunks, and raccoons, as well as providing protection from some predators. Slime molds, often brightly colored, use the nutrients from decaying logs.

Depending on the season and the moisture level, flipping a log may reveal worms, beetles, centipedes, spiders, salamanders, newts, and other life forms. It may also reveal a surprised snake—or a nest of snakes—resting or hiding from predators.

25 *November* | ROOSTING: SOLITARY AND COMMUNAL

After the breeding season, few birds still use their nests for roosting. By this time, nests have lost their comfort and protection and are very likely harboring parasites. Consequently, in searching for a place to sleep, birds adopt several strategies. Ducks and geese remain on water (predators generate vibrations through the water, waking them up) or sleep on the shorelines of small islands. Herons and great egrets roost in large numbers on waterside trees. Large flocks of shorebirds settle in on open beaches and rely on alarms raised by members of the flock. Birds of prey occupy trees. Owls and woodpeckers roost in tree cavities. Grouse and quail are extremely vulnerable in their shrubby or grassy habitats, so they are heavily dependent on protective camouflage, whereas songbirds use dense vegetation to avoid predators.

Some birds change their roosting habits with the season. For example, male red-winged blackbirds usually roost alone in their territories when breeding, but crowd together at night during the rest of the year. In their northern range, both sexes congregate in dense premigration flocks in marshes, as they do at Jug Bay, and then move south, sometimes joined by common grackles and brown-headed cowbirds. Most red-winged blackbird populations in America's interior link up with other species using the Mississippi Flyway and produce the largest concentrations of wintering red-wings on the continent, often roosting together with European starlings and at times becoming a congregation of several million birds.

Small groups of nuthatches or brown creepers spend the night together in tree cavities.

Crows, swifts, and, increasingly, common grackles also adopt communal roosting sites. Such behavior is an attractive proposition, as long as the costs of intensified competition are outweighed by the benefits of increased safety from predators for the older birds, and the advantages of locating rich food supplies override reduced nighttime safety for the young, since they occupy the vulnerable edges of roosts.

Nonetheless, few roosts are completely safe, and some birds have developed the ability to literally sleep with one eye open. Unlike humans, the eyes of most birds send information to only one side of the brain. Unihemispheric slow-wave sleep (USWS) allows birds to send one hemisphere of the brain into a deep sleep, while leaving the other hemisphere awake and alert. Birds can turn USWS on and off, depending on the safety of their roost. For example, when a large flock of ducks is roosting on an open lake, the brains of the birds in the center of the flock may shut down completely, while the more vulnerable birds at the edge of the flock may enter USWS to stay alert. Moreover, scientists such as Gian Mascetti suspect that some birds use USWS to sleep while in flight.

27 November | WHITE-TAILED DEER: IMPACTS OF HUNTS

With shortening days and cooler temperatures, mating for white-tailed deer lasts from late October to late November. With their hormones raging (called estrus in does and rut in bucks), both sexes tend to gather and are more likely to be observed, especially at dawn and dusk. Bucks fight over territory, using their antlers to spar. The antlers of adult males begin to grow an inch a day in late spring and are covered with a highly vascularized tissue known as velvet. The velvet gradually dries out and the antlers harden in the fall, due to calcification stimulated by rising testosterone levels before and during the mating season. The sound of clashing can attract does in heat, and a buck may lose up to 25 percent of his body weight from constantly seeking and chasing does. Antlers are shed in the winter months but are rarely found, since they are rich source of calcium and other nutrients for rodents and other small mammals.

In early spring, does give birth to one to three fawns. They are covered in white spots, can walk in an hour, and search for food a few days later. If a mother has more than one fawn, she will hide them in separate places, where they lie stretched out on the ground. The fawns are weaned after 10 weeks, with females remaining with their mother for two years and males leaving after one year.

The Sanctuary is closed for several days during the months of November, December, and January for deer hunting.[5] The aim is to maintain a healthy deer population in balance with resources in the habitat. Effective management amounts to good stewardship of a valuable resource at a population level that minimizes or eliminates damage to wildflowers, songbirds, and other resources in the ecosystem.

A preliminary study of the Sanctuary by Yasmine Hentati, conducted in 2016, suggests that it houses a relatively large, dense white-tailed deer population and, thus, that there is a need to determine how it will be affected by continued hunts. The study's mention of these deer consuming soybean crops near the Sanctuary in the evenings is especially intriguing, with observed numbers increasing more than threefold once farmlands were included in the survey route. The Sanctuary may be benefiting from their use of these crops, since, if deer are consuming soybeans as their main food source in the summer, they are browsing less often in the forest undergrowth.

The north wind is now sweeping the denuded marsh and shaking leafless branches. Reptiles and amphibians have secured warm hiding places and have settled in for a long sleep. Mammals have donned their warm seasonal coats and begun to build their winter abodes. Fish have fled to deep water. There are no insects to variously annoy or delight, and the landscape is set with gray tones. The consuming fires of autumn have almost finished devouring shrubs and trees. Only beech trees flaunt their last leaves, and their trunks and branches shine in the wintery sun as they arch over clear, plant-free floors.

November brings declining temperatures and shorter days, pressuring all wildlife. Leaf fall lays bare old bird nests and squirrel dreys. A few lingering plants provide food for last minute insects, and wintering avian flocks begin to form. Many animals are hibernating, and plants bare of leaves or fleshy parts have hunkered down for the long haul. The architecture of winter is now upon us.

Low tide in Lower Glebe Marsh in winter, exposing decaying stalks of last summer's emergent vegetation. *Jug Bay Wetlands Sanctuary*

DECEMBER
The Long Night Moon

I prefer winter and fall, when you feel the bone structure of the landscape—
the loneliness of it, the dead feeling of winter. Something waits beneath it, the
whole story doesn't show.

Andrew Wyeth, interview in *Life* magazine

1 December | BLACK DEATH

It is in death's repose. Its magnificent heavy bill and shiny black feathers are
being taken to earth. The hefty claws are fast closed and the eyes glazed.
Bearing no old feathers, it was clearly in the prime of life. There is no obvious
explanation for its death: no marks, no suggestions of foul play. And I would not
be surprised if it took wing, so entire is it unto itself.

I draw closer and gently turn over the body of an American crow. No predator
has arrived, and the body is slightly warm, in part because of the mocking sun.
Wing feathers spread like fingers, as if in supplication.

A while ago it had all of nature to explore. It roved across spreading meadows,
winding rivers, and forests aflame with the orange and gold of autumn. On high,
the crow may have been caught in drifting shadows cast by fleecy clouds beneath
a blue sky. It would have cawed at red barns and school buses, and joined raucous
flocks foraging in fields or inspected a landfill to plunder garbage to ill effect.

Last night before darkness, the now lifeless crow doubtless assembled with
its kind, calling, chasing, and perhaps fighting, and then headed along familiar
flight lines to a communal roost. There, it would have exchanged information on
sources of food and enjoyed protection from predators and the elements. At dawn,
it greeted the rising sun for the last time.

I leave, wondering when predators will find this treasure and recycle its parts—
atoms, molecules, and the like—replenishing the earth and taking on new life in
a harvest a year from now.

3 *December* | Seeing Red: The Perigee Moon

The supermoon phenomenon happens when the moon reaches its closest point to Earth, and on this night, as it rises above the horizon, its red glow is spectacular. It is also a full moon, its brighter light illuminating a wonderful panorama of the marsh.

I have already seen white-tailed deer close to the road on my way to Jug Bay and know the effects of moonlight on wildlife are various, with many species adjusting their behavior in keeping with the lunar cycle.[1] It may provide a time cue for reproduction (as in the case of an intensive breeding period for amphibians), affect the use of the senses (for communication, navigation, and prey or predator location), or indirectly modify the activities of predators, competitors, and prey.

Increased departures during the fall migration of land-based birds are correlated with increased lunar light. Whip-poor-wills tend to synchronize their nesting cycle with the lunar cycle, with hatching making the highest energy demand on the nestlings, coinciding with periods having the most moonlight. Eurasian eagle owls use their white throat feathers during vocal displays to communicate with other birds during a full moon, when their feathers are more visible. Other owls, however, reduce their activity during this time to avoid predators—a widely adopted strategy across the animal kingdom. Rodents are preyed on by owls during full-moon nights, so many species of these mammals curtail their actions or shift them to more sheltered habitats.

Should the increase in foraging success outweigh predation risk, a higher activity level during moonlit nights is expected. For example, the larvae of some dragonfly-like insects dig larger holes for trapping insect prey during full moons, anticipating that the extra effort will pay off.

Many of the above-mentioned instances involve internal clocks. In some, circalunar clocks control monthly rhythms. In other cases, circadian (24-hour) clocks respond to moonlight, and daily activity phases are shifted accordingly. The mechanisms and photoreceptors used to detect moonlight and entrain or shift the clock, however, have been revealed for only a few organisms.

Chronobiology involves the study of biological rhythms, and researchers in moonlight chronobiology—such as Noga Kronfeld-Schor and colleagues—have begun to investigate the interference of artificial light in the functioning of internally derived cycles. Initial studies have found that some nocturnal animals are less likely to forage and follow other natural patterns of activity under such conditions.

Two other supermoons are expected to occur during this particular winter: one on 2 January and another on 31 January (a blue moon).

5 *December* | A Spotted Turtle's Hibernation

I choose to follow a lone spotted turtle, as it may well be the last I will see before next spring. It's on the woodland's edge, and a favorable incline helps it rustle through piled leaves as it heads for the marsh. Sometimes a log requires a detour, and frequent flops in sunken leaves make the journey a trifle noisy, as well as visible. But it is keeping an eye out for any predators or disturbance, and I am careful to keep my distance. Occasionally the turtle rests in patches fully open to the sunlight—its last basking until stirred by a warming spring. On it goes, following easy contours and moist passageways. The pace quickens when a North American river otter comes within reach, and the turtle has no hesitation in striking out for the waterline. With what seems a dash, it plunges quickly into shaded cold water and disappears from view.

But before I bid it farewell, the turtle breaks the silvery surface, looks around, and then plops back. It is descending into deeper water, perhaps having taken its last surface breath of the season. There, a different world unfolds as it follows subterranean contours. It cannot hear the passing Canada geese, or the scolding Carolina chickadees, or see the setting sun. In this cold new world, it must soon dig into the mud and vegetation of slowly moving water to secure a firm lodging for the next five months. Such a hibernaculum has to be deep enough to not freeze completely, but shallow enough to thaw quickly in the spring.

This spotted turtle may find a place to hibernate in congregations (up to 12 individuals have been discovered) and thus be safe enough to avoid being killed by a muskrat. Let it sleep for now.

White perch are starting to gather over shell-strewn bottoms of the deeper waters in Chesapeake Bay.

6 *December* | Plants: Alive, Alive-O!

Once shed of summer's green and autumn's multicolored hues, trees seem somber and lifeless to us, but nothing could be further from the truth. A leafless tree in fall and winter is close to popping with life, and it is merely holding its vegetative breath. Nestled within its buds are thousands of miniature, fully formed infant leaves, foliage the tree will finally exhale in the great green gush of spring.

Like the winter buds of trees, herbs have their own means of getting a head start for the coming spring. The dried remains of some plants, such as common mulleins, common evening primroses, purpleleaf willowherbs, and plantains, are connected to next year's growth in the form of a basal rosette, with green leaves clustered next to the ground. They hide during the cold winter months, awaiting warm days, when the leaves will shoot upward.

Another form of protection during winter occurs along a stem when its internodes are shortened, bringing the leaves closer together, as in wild lettuces and some succulents.[2] In certain instances, the rosette persists at the base of the plant and is connected to a taproot, such as in common dandelions.

The harshness and cold air of wintery times ahead suggest a dull season for wildlife, but an avid eye can capture a richness that does not imbue our landscapes in warmer months. Such has been the case today.

9 December | First Snow of Winter and the Subnivium

A steady fall of snow, beginning in midmorning, has created a wintry landscape, putting all wildlife on notice to glean food for the cold days and nights ahead. Sparrows and juncos are rumbling in the fallen leaves, and chickadees, tufted titmice, and nuthatches inspect bark for possible morsels. The gray, sullen waters of the river and marsh are stilled at the turning of the tide, with snow gathering on cattails, relieving the gloom. With the predicted two to three inches of snow, animals will require more food to keep warm, forcing them to take greater risks. This means that today may prove to be one of the best times to see predators, and I keep an eye out for raptors in search of a meal.

For plants and animals facing harsh winters, blankets of snow can provide a secure refuge from biting winds and subzero temperatures. But in a climate-changing world, snow cover in the Northern Hemisphere is in decline, placing at risk those terrestrial organisms dependent on the space beneath the snow to survive.

In a 2013 report, Jonathan Pauli and a team of fellow scientists from the University of Wisconsin–Madison describe the gradual decay of the subnivium (the seasonal microenvironment beneath the snow), a habitat where life forms—from microbes to mammals—take full advantage of warmer temperatures, near constant humidity, and the absence of wind. The snow retains heat radiating from the ground, allows plants to photosynthesize, and offers a haven for insects, reptiles, amphibians, and many other organisms.

Between 1970 and 2010, snow cover in the Northern Hemisphere (containing the largest land masses affected by snow) has diminished by 0.3 million square miles per decade during the critical spring months of March and April. Maximum snow cover has moved from February to January, and spring melt has accelerated by almost two weeks. Consequently, thresholds are being breached, beyond which some organisms may be unable to survive. Reptiles and amphibians will be put at risk when fluctuating temperatures bring them prematurely out of their winter torpor, only to be pounded by late spring storms or significant drops in temperatures. Insects will also suffer, and migrating birds may find resources scarce when the protective snow cover is absent or sparse. Voles and shrews thriving in networks of tunnels in the subnivium would experience a loss of their snowy refuge and face greater metabolic demands to cope with more-frequent and harsher exposure to the elements. Vegetation directly experiencing cold temperatures and a greater number of freeze and thaw cycles would suffer tissue damage below and above ground, resulting in higher plant mortality, delayed flowering, and reduced biomass.

The authors of the report conclude that these climatic changes will most likely modify the defining qualities of the subnivium, resulting in broad-scale shifts to distributions of species that are dependent on these refugia. Resultant changes to the subnivium, however, will be spatially and temporally variable. These scientists believe that ecologists and managers are overlooking this widespread, crucial, and vulnerable seasonal refugium—rapidly deteriorating, due to global climate change.

14 December | WHISTLING AND CLAPPING WINGS

Many of us are familiar with the whistling wings of a mourning dove, and its suddenness during takeoff and landing, but how often do we ponder the uniqueness of this sound? Explanations are much debated, with the latest thinking suggesting it may be a survival tactic involving the anatomy of the dove's wings.

As an article by Simon Foden notes, the ends of a mourning dove's flight feathers are contoured and create an audible, high-pitched vibration when the wings flutter rapidly, which may serve as a warning to neighboring birds. Foden's studies, conducted at Texas A&M University, have found that mourning doves are more alert after being exposed to a recording of the whistling sound compared with recordings of their calls and chatter. Furthermore, as a mourning dove has a limited song range, it relies on wing sounds for communications with other birds in the flock. These make nest calls and partake in "cooing," but they are capable of very few other vocalizations.

Another study, by Mae Hingee and Robert Magrath at the Australian National University, using crested doves, concluded that it is the faster tempo of the wing whistle when the birds are startled, not loudness, that signals alarm. The authors attribute the sound to a narrow outer feather on the wing.

A mourning dove's wings also play an important role in courtship, involving a flap-glide-flight display. After rising from a perch and making several deep up-and-down wing beats, with feathers hitting to make a flapping sound as they touch under the body, the bird dives in a long, often-spiraling glide, with slightly down-turned wings.

15 December | Snow Language and Relationships

Another dusting of snow, and my mind turns to recalling how ecologists studying winter have adopted many of the terms used by North American Indians to characterize the types of snow we encounter. For example, "quali" describes falling snow that collects on tree branches, "api" for snow on the ground, and "siqoq" for drifting snow. Current classification defines three general types: falling snow (precipitation), snow on the ground (unmetamorphosed or metamorphosed), and surface-generated ice features, the latter broken down into hoar, crusts, and runoff channels.

Snowflakes falling to earth consist of one or more ice crystals, their type determined by temperature and water vapor. They include plates, stellar crystals, columns, needles, spatial dendrites (a branching jumble of individual crystals), and capped columns. Columns are hollow, needles are solid, and spatial dendrites are three dimensional.

Scientists are beginning to understand relationships between falling snow and living organisms. For example, spatial dendrites contribute to snow resting on branches more readily than other crystals, which spares lichens and mosses. Snow on needles may cause an avalanche, and riming (ice crystals growing in the air) may overload branches, causing breakage, or perhaps coat animals, sometimes causing their death.

An example of the impact of snow's surface-generated features on wildlife is that of runoff channels, formed either during warm spells or in the spring, when melting begins. The snow surface develops shallow, somewhat parallel troughs aligned above the surface water. During very warm spells, meltwater percolates down through the snowpack, drenching the creatures below, who may die of hypothermia or drown.

On this day, snow mixed with snow pellets (also known as graupel, or soft hail)[3] fell across the region, and I wonder about its impacts on the Sanctuary's wildlife. It may have frustrated sparrows grubbing in the woodland floor, but it quite possibly concealed the movements of mice from their predators, as well as insulated plants and small animals from the chilling wind. Birches will bend, and conifers can catch enough frigid precipitation to secure a warmer, less snow-filled place below.

Wildlife adapts to the character of the snow, and snow shapes nature on a larger scale. By protecting insects and small mammals from ground-feeding predators and greening plants from browsers, it enhances their populations and their diversity. Snow-generated food limitations create an opposite effect. Thus, snow may well determine life throughout the year.

17 *December* | BROWN CREEPER: GOES TO THE BOTTOM OF THINGS

Brown creepers must be one of the most hyperactive foragers. Having flown to the bottom of a tree, a tiny, mouse-like, long-tailed brown creeper climbs up in short, jerky movements, often disappearing on the far side, as if to play hide-and-seek. After reaching the top, it creeps along one of the largest branches before fluttering down to the base of another trunk to begin the routine all over again. Not surprisingly, these birds have been referred to as (seemingly) legless, perpetual-motion bark wrens.

The ornithologist Arthur Cleveland Bent described this sole native species of tree creeper perfectly: "The Brown Creeper, as he hitches along the bole of a tree, looks like a fragment of detached bark that is defying the law of gravitation

A brown creeper. *Frode Jacobsen*

by moving upward over the trunk, and as he flies off to another tree he resembles a little dry leaf blown about by the wind."

Unlike nuthatches and woodpeckers, creepers never move head down on any portion of the tree, but they can back down tail first and easily creep along the underside of a branch. Birders find that this species is especially attracted to wooded swampland, with its greater frequency of dead trees with hanging strips of loose bark.

During winter, brown creepers may band with chickadees, nuthatches, and woodpeckers as they flock in search of food under collective protection. Moreover, the creepers' cryptic plumage merges with the bark of most trees, so remaining motionless helps them escape detection.

Brown creepers are insectivorous, probing tree trunks and limbs with their slim, down-curved bills, their tails braced against the bark. A recent paper by Jan Reese, reporting a postmortem on a creeper, found an object consistent with an eroded lead shot in the gizzard and physiological signs commonly associated with lead poisoning. Possibly the creeper ingested the pellet while searching for food in tree bark. If so, other bark gleaners—such as woodpeckers, nuthatches, and various warbler species—may also be at risk.

19 *December* | Beavers (No Song, but Smell) and Muskrats Munching in the Marsh

I stand on a small platform overlooking Upper Glebe Marsh and hear the cattails rattle before the cold wind. Sunlight plays on their gently swaying brown flower heads, sentinels in the wind, with the water carrying their fluffy seeds far and wide.[4] The river reflects the icy blue sky, and a passing skein of Canada geese land with enfolded wings and start to cruise upriver in disciplined order, their watery wakes turning the reflections of trees and the sun into a swirl of color. I sit back and, though the wind makes my eyes tear, I admire the muted tones of the marsh and the river: the grays and browns, blues and whites, and silvery ripples pushed by the wind.

I walk to the Beaver Pond to check on the North American beavers and their preparations for winter. They have been busy this autumn, reconstructing the dam breached during a storm, repairing their home, and establishing caches.[5] Trees and other vegetation around the pond continue to be a source of food and construction materials for them. The felled and gnawed broadleaved trees will regrow as a coppice, providing easy-to-reach stems and leaves for food in subsequent years, while those trees and shrubs that the beavers killed—by inundation as the water built up and then by drowning in the filled pond—have created standing dead wood, a microhabitat for many plants and animals.

Beavers have large, sharp upper and lower incisors with which to cut trees and peel fresh bark (which they eat). Incisors continue to grow during a beaver's lifetime, as these teeth are worn down by tree cutting and feeding. Extra digestive glands, and a side branch off the intestines harboring special microorganisms, enhance digestion. A beaver's ability to digest cellulose, however, is no greater than other mammals that don't chew their cud, so it must ingest large amounts of fibrous woody material for adequate nutrition.

During dam reconstruction, peeled logs and branches were carried to the dam and placed parallel to the water current, with mud and stones positioned on the upside of the logs until the sound of flowing water was abated or minimized and a pond was created. The lodge, with an underwater entrance denying most animals access, is in the dam's center. This home normally has two dens, one for drying off and the other for where the family lives. A vent hole at the top of the lodge allows

A snow-covered North American beaver lodge in winter, visible above the ice-covered Beaver Pond. *Rob McEachern*

an exchange of oxygen, carbon dioxide, and other gases. Two or three exits are built, so family members can have more than one way to enter the surrounding water.

As well as providing a secure home, the lodge helps ensure all occupants have a relatively warm microclimate in which to survive the winter. Indeed, plumes of warm air have been observed exiting the vent of the lodge. Beavers usually roam until just before autumn, when they return to their old lodges and gather their winter stock of wood into caches, usually taking the form of a raft of nonfood or low preference items (e.g., alder logs) and preferred foods (e.g., maples and tulip poplars, or tuliptrees).

Other ways beavers cope with seasonal changes involve reducing their metabolism and storing fat (up to 30–40 percent of their body weight) during autumn. All of the adaptions mentioned here have allowed these remarkable animals to occupy a wide range of aquatic habitats, extending from Alaska to Mexico.

The basic units of beaver social organization are families, consisting of a monogamous pair of adults and their kits and yearlings. Beavers mate between January and March, and litters of one to nine kits (averaging three to four) are born between April and June. They remain in the lodge for a month, leaving it to swim and search for food, which usually is softer summer fare. Lactation may continue for some 90 days, but kits are usually weaned in two weeks. Most will remain with the adults until they are almost two years old. They then go in search of mates and suitable spots to begin their own colonies, often several miles away.

Unlike primarily terrestrial vertebrates, such as songbirds or squirrels, beavers are nocturnal or crepuscular. Consequently, songs and visual displays are not effective means of communicating to other beavers that a territory is occupied. Thus beavers are diligent in maintaining and defending their territories by marking them with scent mounds, sprayed with a mixture of urine and chemicals produced by castor sacs located between the pelvis and the base of the tail.[6] Because they invest so much energy in their territories, beavers are intolerant of intruders, and the holder of the territory is more likely to escalate any trespass into an aggressive encounter.

Most of us are familiar with the tail slap used by adult females to signal danger and to startle or frighten enemies when kits are resident in the lodge. Beavers also communicate, however, via whines and growls and hisses.

Although closely related to voles, muskrats are more beaver-like in their lifestyle. As marsh vegetation dies back, their homes (pushups), consisting of domes of cattails, become visible, often rising two to three feet above the water. Like beavers, a muskrat builds underwater entrances and tunnels to enter its pushup, which houses a nesting chamber, or den. It often builds two other structures: a feeding station (looking like a small pushup), and a scent post of matgrass and mud, where the male marks his territory using musk from his scent glands. Muskrats do not store food for the winter, but sometimes they eat the inside of their pushups.

Richard Yahner's studies on the population cycle of muskrats have concluded that marshes and other aquatic habitats occupied by high numbers of muskrats can be severely degraded by an "eat out" with their consuming the shoots of plants in the spring, leaves and stems in the summer, and roots throughout the winter. Consequently, a particular challenge for wildlife professionals interested in managing both waterfowl and muskrat populations is to maintain a hemimarsh, consisting of equal portions of emergent vegetation and open water.

The Beaver's Tail

The flat tail of a large North American beaver is covered with leathery scales and sparse, coarse hairs roughly 15 inches long and 6 inches wide.

In the water, the beavers use their flexible tails as a four-way rudder when swimming. After being disturbed, a beaver will slap the water loudly with its tail, alerting all other beavers in the vicinity of danger and possibly frightening potential predators.

On land, the tail acts as a prop when the beavers are sitting or standing upright. It also serves as a counterbalance and support when these animals are walking on their hind legs to carry building materials with their teeth, front legs, and paws.

During winter, the tail is used to store fat. Because it is almost hairless, it releases body heat, helping beavers regulate their body temperature in summertime.

Like beavers, muskrats can modify aquatic environments and change the abundance and diversity of plant life, but—unlike beavers, with their dramatic population fluctuations—muskrats emulate their vole relatives by maintaining stable numbers.

21 December | WINTER SOLSTICE

The winter solstice marks the return of light to our skies and occurs when the North Pole is at its greatest tilt away from the sun. Between the summer solstice and now, days have grown shorter, and the shallower angles of sunlight have cast long shadows across the landscape. The winter solstice marks the moment the sun, which rises just south of east and sets south of west, shines at its most southern point, directly over the Tropic of Capricorn.

In North America and some other countries in the Northern Hemisphere, the solstice marks the first day of winter. The official date varies, however, depending on a country's climate and whether it follows astronomical or meteorological seasons.

All day, clouds competed with the sun, finally filtering it before the afternoon ceded to the earliest of dusks. Tomorrow a reversal will begin, and the season will have turned.

26 *December* | THE MARSH IN WINTER

Out on the marsh, blue water shows through shifting ice.

Tall brown reeds, slim as dancers, bend in the breeze.

A hundred thousand cattails, each one lit

by the low-angled light of a westering sun,

each brown seed head blazing

like the head of a saint.

Timothy Walsh, "The Marsh in Winter"

Stillness beneath the water is illusory, even when ice forms and is covered with snow. Submerged plants and other forms of aquatic life continue to photo-synthesize (barely) in temperatures just above freezing. On occasion, prolonged poor light reduces oxygen levels dramatically and imperils survival. To counter this, aquatic organisms endure as dormant seeds and eggs, or by lowering their metabolic rates.

I edge into the marsh, with its partly dead vegetation and ice, then plunge a net in the detrital ooze and lift out a winter storehouse. My fingers provide some warmth, and, slowly, sluggish aquatic bugs, tadpoles, and the imagoes of dragon-flies, damselflies, and mayflies reveal themselves. I wade deeper and do another sweep. This time I pull out some of the same insects, as well as a small bluegill in a very torpid state. Amid the detritus and mud, all await more congenial tempera-tures, and I quickly return them to their winter habitat before they suffer a too rapid revival.

I move into the deep mucky bottom of dense vegetation and am lucky to net a hibernating spotted (more like a "polka-dot") turtle. Perhaps it's the one I saw earlier this month, and I ponder how it survives low oxygen levels. Some herpe-tologists believe these turtles absorb some oxygen directly through their exposed skin and neutralize accumulated lactic acid with chemical buffers.

Turtles will sometimes awaken under the ice, as observed by Michael Caduto as he "ventured out onto the ice where, eerily suspended above the living diorama," he saw a spotted turtle swim lazily by.

A very low tide later in the day reveals expansive mud and a vast biomass of dead marsh plants (often appearing as muck). Though we may pass over this landscape

Bottoming Out: Death and Decay of Marsh Plants

Research conducted at the Jug Bay marsh in the winter of 1990 by Marilyn Fogel and Noreen Tuross investigated what happened to dead plant material (detritus) when it finally collapses onto the mud and undergoes bacterial decomposition and degradation.

After collecting samples of leaves and roots of wild rice, spatterdocks, American hollies, smooth alders, mountain laurels, oaks, and loblolly pines when the plants were dying back, the research workers sewed the material into fine-mesh nylon bags and placed them on the marsh surface, as well as in the mud. Several months or years later, the bags were removed and the material inside was analyzed in the laboratory. Weight loss indicated the rate of overall decay, and changes in amino acids and isotopic forms (i.e., particular chemical forms) of carbon and nitrogen helped to determine what happened to the basic plant elements. An analysis employing mass spectrometry and antibodies determined whether the original plant biochemicals remained, and another using enzymes described the protein content before and after decomposition in the mud.

Their research found that following one to two years of decomposition, between 10 and 40 percent of the plant material remained intact, with wild rice decomposing more slowly than spatterdocks. After the plants fell into the mud, microbes attacked the tissues fairly quickly, with carbohydrate-rich roots and stalks proving to be more palatable to microbes, thus losing weight more quickly than leaves.

Over time, microbes rework and replace the original plant compounds with new combinations and, in as little as three months, turn plant proteins into permanent organic mud, imparting a dark brown color to it. The decayed plants become the source of gases bubbling up through marsh sediments and nutrients cycling in the tidal waters.

The scientists noted the importance of decomposition in sustaining the Jug Bay wetlands: without decay, the Patuxent River Valley would be filled up with

A flock of winter waterfowl foraging amid plant material bottoming out
in Upper Glebe Marsh. *Rob McEachern*

dead vegetation to the level of the Route 4 bridge. They also noted that while most
animals don't eat dead plants (save for red-bellied and eastern painted turtles),
many small invertebrates feast on the bacteria and fungi. These invertebrates are
eaten by small fish and crabs, which, in turn, are eaten by bigger fish and ospreys,
passing energy up the food chain. Meanwhile, nutrients and minerals are being
released into the water.

Wetlands are said to improve water quality by removing overabundant nutrients, such as nitrogen and phosphorus, but how true is that of Jug Bay if such
nutrients are stored only during the growing season? Nonetheless, when plants
die back in the fall, their remains cover the upper sediments, burying heavy metals
much more efficiently than in other types of wetlands.

with little thought, such decomposition fuels the water with nutrients, permitting the rapid growth of plankton in the spring.

28 December | Winter Ecology

I find it most enlightening to read Peter Marchand's *Life in the Cold*. He suggests that two basic strategies have been adopted to survive the rigors of winter: avoidance (migration) and confrontation (hibernation and resistance). Though migration would seem to be a preferred strategy, it is precarious at best, with its demand for energy, avoidance of predators (including hunters), climatic vagaries, and habitat erasure. Even hibernators court risks, as they generally cannot let their body temperatures fall below freezing. When threatened, they must rouse from sleep and restore their body temperature to normal before resuming hibernation. This means metabolizing considerable amounts of fat, so it is not surprising that hibernation is predominantly a midlatitude adaptation.

For amphibians and reptiles, their options are limited to hibernation. Frogs and toads seek underground burrows (which often are very deep), occupy streambanks (where temperatures are less likely to drop below freezing), or bury themselves in the soft mud of marshes or other aquatic bodies (frogs can survive temperatures that are a few degrees below freezing). Reptiles also search for safe havens under decaying logs, in crevices, or in the burrows of other wintering animals, often evicting the previous occupants.

For many organisms, winter proves to be an endurance test: insect larvae and pupae use biochemical mechanisms to survive freezing temperatures, and some birds, such as chickadees, undergo nightly torpor, or controlled hypothermia, to conserve energy and often adopt communal roosting. Some mammals erect the hairs in their fur, and birds fluff their feathers (piloerection) to enhance the insulative value of trapped air. Many small mammals that are solitary in summer construct communal nests under the snow, and two or more beavers may occupy a lodge, where the central chamber stays well above freezing, even when outside temperatures are very low.

Aquatic invertebrates, under the ice cover, endure in a state of semihibernation—lethargic and slow breathing, but able to respond to physical stimuli. Zooplankton remain suspended in the water column, aided by the higher density of cold water.

A coating of ice actually protects aquatic life from winter's extremes, enabling most to continue in an environment with temperatures just above freezing. Cold-blooded vertebrates decrease their vulnerability by lowering their heartbeats, circulation, and metabolism. Some fish, such as bluegills, remain in a state of suspended animation. Tadpoles, mummichogs, and carnivorous insects (in both adult and immature stages) wend their way and can even survive freezing by burrowing in wetland mud. Watersnakes hibernate in burrows shared with amphibians, while dormant turtles occupy muddy areas just below the waterline, in oxygen-poor conditions.

Phytoplankton photosynthesize throughout the winter, shifting as light levels change under the ice and snow cover. One submerged aquatic plant, pondweed, is known to increase its biomass by active photosynthesis, thereby elevating the amount of dissolved oxygen in the water and inviting a daytime in-migration of some organisms.

Other plants have also evolved means for avoiding the hardships of winter. Seasonal shedding is an obvious case, being a strategy to reduce water loss at a time of stress. Some vegetation overwinters in the form of seeds or rootstocks, and some have evolved morphological adaptations, but most rely on physiological mechanisms, such as acclimation or resistance to freezing.

29 December | ALL IN A TRICE

As I make my way along the Railroad Bed Trail, listening to the creaking and groaning of trees, I notice a male Cooper's hawk eying bushes ahead of me. I stare hard but, for the life of me, cannot make out what has caught his interest. Then he swoops down, plunging through a thicket by sheer force and arching back in one exquisite movement, to alight on a perch, as would a flycatcher. The prey is now visible, bulging from his talons.

As the hawk begins to tear at its prize, reddish feathers start to fall, then grayish-brown ones, which slowly provide clues as to what has been caught. It is a female northern cardinal, being adroitly dismembered by beak and yellow-scaled legs.

I continue to watch, noticing that all life has fallen silent, as if terror is stalking the landscape. After some 10 minutes, the hawk has finished his feast and, with much grace, leaps from limb to limb before taking off for woodland on the other side of the Patuxent River. Killing is a part of nature's daily round.

Goldfinches in Winter

Studies on American goldfinches provide a model that demonstrates how birds keep warm in winter. A summer-caught goldfinch placed in a severely cold locale can produce enough heat to maintain its body temperature for only one hour. A winter-caught bird under identical conditions can keep warm for six to eight hours. The summer-caught bird depletes its carbohydrate stores very quickly and is left with insufficient enzymes to burn fat at a great rate. The winter-caught bird has a different enzyme composition and is able to burn much more fat in relation to its carbohydrates. Consequently, these daytime feeders are able to survive winter nights when they are unable to forage for food.

American goldfinches use many other strategies to battle hostile conditions. As winter approaches, they almost double the number of feathers covering their bodies and, during this season, occupy protective roosts and forage for shorter periods. Such behavior results in roughly a 30 percent savings in energy. At night, their body temperatures may drop, reducing metabolic costs by 20 percent. Under certain conditions, though, goldfinches may shiver all night long.

Source: Dawson and Carey 1976.

The powerful and swift flight of the Cooper's lent it the earlier common names of big blue darter, big stub-winged bullet-hawk, and privateer. But perhaps the most deserved was chicken hawk, in recognition of this bird of prey hanging around poultry yards and being able to carry off pullets almost as heavy as itself.

30 December | ARCTIC BLAST

A strong Arctic air mass is spread over the eastern two-thirds of the continent ahead of the last day of 2017. In the Washington, DC, area, temperatures remain in the low 20s and, when combined with gusty northwesterly winds (15–25 mph), create single digit wind chills. This makes it the coldest December day in the area since 24 December 1989.

As the old year cedes to the new, nighttime temperatures are expected to range from 7°F to 12°F and 10–15 mph winds from the northwest will keep the wind chills at or below zero.

31 *December* | ON THE CUSP OF THE NEW YEAR

Winter is the anvil on which nature hammers out next spring. Its furnace is cold fire. It fashions motes of life. Even in the utmost extremes of landscape and weather they endure.

Jim Crumley, *The Nature of Winter*

In early morning, from the Observation Deck overlooking the Patuxent River, much of the landscape presents a screen of trees against a blue-gray sky flecked with pink and golden rays pulsating in the occasional snow flurries. Below is a khaki-colored grassland easing from woodland and morphing into shoreland, marsh, and dark open water. All is rendered with a limited palette.

As the day progresses, I see a bald eagle soaring overhead, lit from beneath by a watery sun, admire long wakes of glinting ripples that follow a flock of Canada geese cruising in the river, and hear the tiny feet of a white-breasted nuthatch scratching the bark as it hops down a tree. The dark green, leathery fronds of Christmas ferns and American holly trees weighed down with berries lend cheer to the woodland. Though standing plants (weeds, to some) appear withered, brown, and lifeless, they continue to disperse seeds from the tips of their stalks or remain alive within their roots. In the marsh, winter birds in search of moth caterpillars are disturbing the sprinkling of snow on cattail heads, and the showering flakes become iridescent in the sunlight.

In the afternoon, thousands of ring-billed and herring gulls assemble in the marsh—a way station in the daily commute between their lunch spot at a landfill in Prince George's County and their bedroom community on the waters of Chesapeake Bay, to the east. Lora Wondolowski, a graduate student, found that gulls, after consuming garbage, bathe, preen, and drink in the marsh, while also keeping watch for predatory bald eagles. Apparently, younger gulls in the area linger longer into spring than do the adults, who need to fly north and west to their breeding grounds.

Later, the fading orange glow of the setting sun against the blue-gray sky merges with the darkening landscape and suddenly becomes one as I leave the Sanctuary. The calls of Canada geese in skeins arching high above the marsh will carry these and other memories into the New Year and beyond.

I leave as the inexorable tide turns. Though the sun has set and all is stilled, a new cycle is stirring and all life will awaken to another season. As the Roman orator Symmachus has said, "We see the same stars, the sky is shared by all, the same world surrounds us." None of us follows the same route in relating to the natural world; but all of us are irrevocably tied to its wonder.

December has its short days in the Sanctuary's leafless and watery terrain punctuated with moments of beauty, held in an icy clasp. The area's wild skies, where birds rise and fall over land sugared with the first snow, are sometimes hidden by fog that slowly dissolves before the rising sun. As always, Carolina wrens are ever ready to sing the old year away and greet the new.

Epilogue

Those who dwell, as scientists or laymen, among the beauties and mysteries of the earth, are never alone or weary of life. Whatever the vexations or concerns of their personal lives, their thoughts can find paths that lead to inner contentment and to renewed excitement in living.

Rachel Carson, *The Sense of Wonder*

In nature's infinite book of secrecy
A little I can read.

William Shakespeare, *Anthony and Cleopatra*, act 1, scene 2

Throughout one year, I have been privileged to witness the passing seasons in but a tiny corner of the natural world. I often felt a strange sense of humility and pride, emotions I had hitherto thought irreconcilable. Beauty and grandeur were my most constant companions, and there were occasions when I lost myself pondering a creature, the fleeting moods of the day, or, as Darwin put it in *On the Origin of Species*, "a tangled bank, clothed with many plants of many kinds, with birds singing on the bushes, with various insects flitting about, and with worms crawling through the damp earth."

The diary format kept me disciplined, as well as aware that I was observing an ever-changing and challenging cycle. And though birds are my major passion, I viewed them through the lens of a naturalist, thereby focusing on all wildlife. This stretched my knowledge and understanding of the natural world and brought insights that delighted as much as intrigued. Events often piled up, sometimes competing for space, and I had to be both judicious and economical in maintaining the story line.

Many accounts of the passage of the seasons are light in recording rainfall, storms, and varying temperatures, so I found myself paying heed to changing weather conditions by day and by month. Climate change is the joker in the pack, and there were dramatic departures, breaking records and having consequent impacts on wildlife and the landscape. Hot and cold days put nature on a roller coaster, sometimes with a disparity of 20 or more degrees in one day. The average temperature for the year was 60°F, 2.6 degrees above normal, making it the second-warmest year since 1871, with February and April as the warmest recorded months, October as the third warmest, and June as the tenth warmest. A record low of 26°F was recorded on 11 November.[1] Total precipitation was 32.3 inches, 10 inches less than normal.

The above is only a snapshot of climate within one year. What will climate change hold at the end of this century for the Jug Bay Wetlands Sanctuary, and Jug Bay itself? Warmer air and water temperatures, rising waters, increased precipitation, elevated nutrient and sediment levels, and a scarcity of dissolved oxygen are likely to exclude some species, and even habitats, and favor others. Bacteria and other cold-blooded organisms will consume more oxygen, because of their increased metabolic rates in warmer waters. Greater stream flows will dump higher quantities of nutrients and sediments in bodies of water, with the former stimulating plant production and the latter clouding the water, thereby reducing the amount of light reaching submerged aquatic vegetation. With higher carbon dioxide levels available to plants, more carbon compounds could end up in wetland soils and more could be emitted back into the air as methane. The growth of trees, sedges, and emergent macrophytes in the wetlands may benefit from greater carbon dioxide levels; grasses, including wild rice, are unlikely to respond. Thus the species composition of the Sanctuary would be tilted toward emergent aquatic macrophytes such as spatterdocks, pickerelweeds, green arrow arums, and arrowheads (assuming that they are not submerged). Steeply sloping banks will limit the migration of wetland plants inland, rising waters will shrink the tidal marshes, and wetland birds would thus lose intertidal habitat. Lastly, sustained salinity spikes may see a decrease, and eventual loss, of species that are critical in maintaining relationships within and the structure of an ecological community (keystone species). In all, both Jug Bay and the Sanctuary will undergo a profound change.

This sword of Damocles encouraged me to look more closely at what lies unseen: the slow but inexorable march of time; the seemingly small changes that mount up; and the many injurious effects on wildlife and, ultimately, our welfare. It made me grasp how much there is to lose. With daily assaults upon what has been gained these past 50 or so years to protect the environment, it also made me wonder what sort of world we will leave for our children and grandchildren. In sum, I believe we are in great danger of holding up a mirror to nature and seeing nothing but ourselves.

So, gentle reader, permit me to say all nature is on your doorstep, waiting to be explored. In this hurried life, it offers us a time to reflect and enjoy this remarkable world before it is too late.

Appendix A | Animals Mentioned in the Text

INVERTEBRATES

Ailanthus webworm moth *Atteva aurea*

Amber snail *Succinea putris*

American ermine moth *Yponomeuta multipunctella*

Blinded sphinx *Paonias excaecatus*

Blue dasher *Pachydiplax longipennis*

Bumblebee *Bombus* sp.

Cabbage white *Pieris rapae*

Carolina ground cricket *Eunemobius carolinus*

Carpenter ant *Camponotus* sp.

Carpenter bee *Xylocopa* sp.

Citrine forktail *Ischnura hastata*

Clouded sulphur *Colias philodice*

Cloudless sulphur *Phoebis sennae*

Common blue damselfly *Enallagma cyathigerum*

Common buckeye *Junonia coenia*

Common whitetail *Plathemis lydia*

Confused ground cricket *Eunemobius confusus*

Crocus geometer moth *Xanthotype sospeta*

Dun skipper *Euphyes vestris*

Eastern comma *Polygonia comma*

Eastern long-tailed skipper *Urbanus proteus*

Eastern tiger swallowtail *Paplio glaucus*

Ebony jewelwing *Calopteryx maculata*

Fairy shrimp *Eubranchipus vernalis*

Fingernail clam *Sphaerium corneum*

Giant diving beetle *Dytiscus marginalis*

Gold moth *Axia* sp.

Grass shrimp *Palaemonetes* sp.

Great spangled fritillary *Speyerta cybele*

Green stink bug *Chinavia hilaris*

Gypsy moth *Lymantria dispar dispar*

Handsome meadow katydid *Orchelimum pulchellum*

Handsome trig *Phyllopalpus pulchellus*

Hay's Spring amphipod *Stygobromus hayi*

Honeybee *Apis* sp.

Hummingbird clearwing *Hemaris thysbe*

Least skipper *Ancyloxypha numitor*

Mole cricket *Neoscapteriscus* sp.

Monarch butterfly *Danaus plexippus*

Mourning cloak *Nymphalis antiopa*

Mustard white *Pieris oleracea*

Nais tiger moth *Apantesis nais*

Orange-tip *Anthocharis cardamines*

Pearl crescent *Phyciodes tharos*

Pink-spotted hawkmoth *Agrius cingulata*

Red admiral *Vanessa atalanta*

Ruby meadowhawk *Sympetrum rubicundulum*

Sac spider *Cheiracanthium* sp.

Silverfish *Lepisma saccharina*

Silvery checkerspot *Chlosyne nycteis*

Six-spotted fishing spider; raft spider *Dolomedes triton*

Sleepy orange *Abaeis nicippe*

Southern flannel moth *Megalopyge opercularis*

Spicebush swallowtail *Papillo troilus*

Spring azure *Celeastrina ladon*

Tinkling ground cricket *Allonemobius tinnulus*

Turbulent phosphila *Phosphila turbulenta*

Water boatman *Micronecta scholtzi*

Water strider; pond skater *Gerrindae* sp.

Whirligig beetle *Gyrinidae* sp.
Zabulon skipper *Poanes zabulon*

Zebra swallowtail *Eurytides marcellus*

VERTEBRATES
Fish

Alewife *Alosa pseudoharengus*
Bluegill *Lepomis macrochirus*
Chain pickerel *Esox niger*
Common carp *Cyprinus carpio*
Gizzard shad *Dorosoma cepedianum*
Killifish *Profundulus* sp.
Largemouth bass *Micropterus salmoides*
Least brook lamprey *Lampetra aepyptera*

Mummichog *Fundulus heteroclitus*
Pumpkinseed sunfish *Lepomis gibbosus*
Redfin pickerel *Esox americanus*
Striped bass *Morone saxatilis*
Tessellated darter *Etheostoma olmstedi*
White crappie *Pomoxis annularis*
White perch *Morone americana*
Yellow perch *Perca flavescens*

Amphibians

American bullfrog *Lithobates catesbeianus*
Eastern American toad *Anaxyrus
    americanus americanus*
Eastern spadefoot *Scaphiopus holbrookii*
Fowler's toad *Anaxyrus fowleri*
Green frog *Lithobates clamitans clamitans*
Green treefrog *Hyla cinera*
Marbled salamander *Ambystoma opacum*

Pacific treefrog *Pseudacris regilla*
Southern leopard frog *Lithobates
    sphenocephalus*
Spotted salamander *Ambystoma
    maculatum*
Spring peeper *Pseudacris crucifer crucifer*
Wood frog *Lithobates sylvaticus*

Reptiles

American mud turtle *Kinosternon* sp.
Common watersnake *Nerodia sipedon*
Eastern box turtle *Terrapene carolina*
Eastern fence lizard *Sceloporus undulatus*
Eastern painted turtle *Chrysemys picta
    picta*

Northern watersnake *Nerodia sipedon
    sipedon*
Red-bellied turtle; northern red-bellied
    cooter *Pseudemys rubriventris*
Snapping turtle *Chelydra serpentina*
Spotted turtle *Clemmys guttata*

Birds

Acadian flycatcher *Empidonax virescens*
Alder flycatcher *Empidonax alnorum*
American bittern *Botaurus lentiginosus*
American black duck *Anas rubripes*
American coot *Fulica americana*
American crow *Corvus brachyrhynchos*

American goldfinch *Spinus tristis*
American redstart *Setophaga ruticilla*
American robin *Turdus migratorius*
American wigeon *Anas americana*
American woodcock *Scolopax minor*
Bald eagle *Haliaeetus leucocephalus*

Baltimore oriole *Icterus galbula*

Barn swallow *Hirundo rustica*

Barred owl *Strix varia*

Bay-breasted warbler *Setophaga castanea*

Belted kingfisher *Megaceryle alcyon*

Black-and-white warbler *Mniotilta varia*

Blackburnian warbler *Setophaga fusca*

Black-capped chickadee *Poecile atricapillus*

Blackpoll warbler *Setophaga striata*

Black-throated blue warbler *Setophaga caerulescens*

Black-throated green warbler *Setophaga virens*

Blue-gray gnatcatcher *Polioptila caerulea*

Blue jay *Cyanocitta cristata*

Blue-winged teal *Anas discors*

Blue-winged warbler *Vermivora cyanoptera*

Bobolink; reedbird *Dolichonyx oryzivorus*

Brown creeper *Certhia americana*

Brown-headed cowbird *Molothrus ater*

Brown thrasher *Toxostoma rufum*

Bufflehead *Bucephala albeola*

Canada goose *Branta canadensis*

Canada warbler *Cardellina canadensis*

Cape May warbler *Setophaga tigrina*

Carolina chickadee *Poecile carolinensis*

Carolina wren *Thryothorus ludovicianus*

Caspian tern *Hydroprogne caspia*

Cerulean warbler *Setophaga cerulea*

Chipping sparrow *Spizella passerina*

Common cuckoo; European cuckoo *Cuculus canorus*

Common grackle *Quiscalus quiscula*

Common poorwill *Phalaenoptilus nuttallii*

Common yellowthroat *Geothlypis trichas*

Cooper's hawk *Accipiter cooperi*

Crested dove; crested pigeon *Ocyphaps lophotes*

Dark-eyed junco *Junco hyemalis*

Double-crested cormorant *Phalacrocorax auritus*

Dowitcher *Limnodromus* sp.

Downy woodpecker *Picoides pubescens*

Eastern bluebird *Sialia sialis*

Eastern kingbird *Tyrannus tyrannus*

Eastern phoebe *Sayornis phoebe*

Eastern towhee *Pipilo erythrophthalmus*

Eastern whip-poor-will *Antrostomus vociferous*

Eastern wood-pewee *Contopus virens*

Eurasian eagle owl *Bubo bubo*

European starling *Sturnus vulgaris*

Field sparrow *Spizella pusilla*

Fish crow *Corvus ossifragus*

Fork-tailed flycatcher *Tyrannus savana*

Forster's tern *Sterna forsteri*

Fox sparrow *Passerella iliaca*

Golden-crowned kinglet *Regulus satrapa*

Gray catbird *Dumetella carolinensis*

Great black-backed gull *Larus marinus*

Great blue heron *Ardea herodias*

Great egret *Ardea alba*

Greater yellowlegs *Tringa melanoleuca*

Great horned owl *Bubo virginianus*

Green heron *Butorides virescens*

Green-winged teal *Anas crecca*

Hairy woodpecker *Picoides villosus*

Hermit thrush *Catharus guttatus*

Herring gull *Larus argentatus*

Hooded merganser *Lophodytes cucullatus*

Hooded warbler *Setophaga citrina*

House finch *Haemorhous mexicanus*

House sparrow *Passer domesticus*

House wren *Troglodytes aedon*

Laughing gull *Leucophaeus atricilla*

Least bittern *Ixobrychus exilis*

Least sandpiper *Calidris minutilla*

Lesser yellowlegs *Tringa flavipes*

Louisiana waterthrush *Parkesia motacilla*

Magnolia warbler *Setophaga magnolia*

Mallard *Anas platyrhynchos*

Marsh wren *Cistothorus palustris*

Merlin *Falco columbiaris*

Mourning dove *Zenaida macroura*

Musician wren *Cyphorhinus arada*

Mute swan *Cygnus olor*

Nashville warbler *Oreothlypis ruficapilla*

Nightingale wren *Microcerculus philomela*

Northern cardinal *Cardinalis cardinalis*

Northern flicker *Colaptes auratus*

Northern harrier *Circus hudsonius*

Northern mockingbird *Mimus polyglottos*

Northern parula *Setophaga americana*

Northern pintail *Anas acuta*

Northern rough-winged swallow *Stelgidopteryx serripennis*

Northern waterthrush *Parkesia noveboracensis*

Orange-crowned warbler *Oreothlypis celata*

Orchard oriole *Icterus spurius*

Osprey *Pandion haliaetus*

Ovenbird *Seiurus aurocapilla*

Palm warbler *Setophaga palmarum*

Pied-billed grebe *Podilymbus podiceps*

Pileated woodpecker *Dryocopus pileatus*

Pine warbler *Setophaga pinus*

Plover *Pluvialis* sp.

Prairie warbler *Setophaga discolor*

Prothonotary warbler; golden swamp warbler *Protonotaria citrea*

Pygmy nuthatch *Sitta pygmaea*

Red-bellied woodpecker *Melanerpes carolinus*

Red-breasted nuthatch *Sitta canadensis*

Red-eyed vireo; preacher bird *Vireo olivaceus*

Red-headed woodpecker *Melanerpes erythrocephalus*

Red knot *Calidris canutus*

Red-shouldered hawk *Buteo lineatus*

Red-tailed hawk *Buteo jamaicensis*

Red-winged blackbird *Agelaius phoeniceus*

Ring-billed gull *Larus delawarensis*

Ring-necked pheasant *Phasianus colchicus*

Ruby-throated hummingbird *Archilochus colubris*

Ruddy turnstone *Arenaria interpres*

Ruffed grouse *Bonasa umbellus*

Rusty blackbird *Euphagus carolinus*

Sanderling *Calidris alba*

Savannah sparrow *Passerculus sandwichensis*

Scarlet tanager *Piranga olivacea*

Semipalmated sandpiper *Calidris pusilla*

Sharp-shinned hawk *Accipiter striatus*

Song sparrow *Melospiza melodia*

Sora; sora rail *Porzana carolina*

Spotted sandpiper *Actitis macularius*

Summer tanager *Piranga rubra*

Swamp sparrow *Melospiza georgiana*

Tennessee warbler *Oreothlypis peregrina*

Tree swallow *Tachycineta bicolor*

Tufted titmouse *Baeolophus bicolor*

Tundra swan; whistling swan *Cygnus columbianus*

Veery *Catharus fuscescens*

Virginia rail *Rallus limicola*

Western bluebird *Sialia mexicana*

White-breasted nuthatch *Sitta carolinensis*

White-crowned sparrow *Zonotrichia leucophrys*

White-eyed vireo *Vireo griseus*

White-throated sparrow *Zonotrichia albicollis*

Wild turkey *Meleagris gallopavo*

Wilson's snipe *Gallinago delicata*

Winter wren *Troglodytes hiemalis*

Wood duck; summer duck *Aix sponsa*

Wood thrush *Hylocichla mustelina*

Worm-eating warbler *Helmitheros vermivorum*

Yellow-bellied sapsucker *Sphyrapicus varius*

Yellow-billed cuckoo *Coccyzus americanus*

Yellow-rumped warbler; myrtle warbler *Setophaga coronata*

Yellow-throated vireo *Vireo flavifrons*

Yellow warbler *Setophaga petechia*

Mammals

American black bear *Ursus americana*

American mink *Neovison vison*

Eastern chipmunk *Tamias striatus*

Eastern cottontail *Sylvilagus floridanus*

Eastern gray squirrel *Sciurus carolinensis*

Eastern red bat *Lasiurus borealis*

Gray fox *Urocyon cinereoargenteus*

Marsh rice rat *Oryzomys palustris*

Meadow vole *Microtus pennsylvanicus*

Moose *Alces alces*

Muskrat *Ondatra zibethicus*

North American beaver *Castor canadensis*

North American river otter *Lontra canadensis*

Northern short-tailed shrew *Blarina brevicauda*

Raccoon *Procyon lotor*

Red fox *Vulpes vulpes*

Vole *Microtus* sp.

White-tailed deer *Odocoileus virginianus*

Appendix B | Plants Mentioned in the Text

FUNGI

Beefsteak polypore *Fistulina hepatica*
Black-footed polypore *Picipes badius*
Coral mushroom *Artomyces pyxidatus*
Fairy-ring mushroom *Marasmius oreades*
Honey mushroom *Armillarias mellea*

Morel *Morchella* sp.
Netted rhodotus *Rhodotus palmatus*
Thin-walled maze polypore; blushing
 bracket *Daedaleopsis confragosa*
Turkey tail *Trametes versicolor*

LICHENS AND MOSSES

Candleflame lichen *Candelaria concolor*
Clubmoss; running pine *Lycopodium
 clavatum*
Earthmoss *Physcomitrella patens*
Gray reindeer lichen *Cladonia rangiferina*
Haircap moss; polytrichum
 moss *Polytrichum commune*
Pincushion moss; leucobryum
 moss *Leucobryum glaucum*
Powdery goldspeck; eggyolk lichen
 Candelariella efflorescens

Prairie sphagnum; blunt-leaved bog
 moss *Sphagnum palustre*
Seastorm lichen *Cetrelia monachorum*
Shield lichen *Parmelia* sp.
Silvergreen bryum moss *Bryum argenteum*
Toothed plagiomnium moss *Plagiomnium
 cuspidatum*
Water moss; brook moss; fountain moss;
 fontinalis moss *Fontinalis* sp.

FERNS

Carolina mosquitofern *Azolla caroliniana*
Christmas fern *Polystichum acrostichoides*
Eastern marsh fern *Thelypteris palustris*

Floating watermoss *Salvinia natans*
Sensitive fern *Onoclea sensibilis*

FLOWERING PLANTS

Alder *Alnus* sp.
American beech *Fagus grandifolia*
American black elderberry *Sambucus
 nigra canadensis*
American holly *Ilex opaca*
American pokeweed *Phytolacca americana*
American strawberry bush; bursting-heart
 Euonymus americanus
American sweetgum *Liquidambar
 styraciflua*

American witchhazel; snapping
 hazel *Hamamelis virginiana*
Arrowleaf tearthumb *Polygonum
 saggitatum*
Arrowwood viburnum; southern
 arrowwood *Viburnum dentatum*
Asiatic tearthumb *Polygonum perfoliatum*
Awlfruit sedge *Carex stipata*
Barnyardgrass *Echinochloa crus-galli*
Basswood *Tilia* sp.

Blackberry *Rubus* sp.

Black cherry *Prunus serotina*

Black chokecherry *Prunus virginiana melanocarpa*

Blackeyed susan *Rudbeckia hirta*

Blackgum *Nyssa sylvatica*

Blackhaw; blackhaw viburnum *Viburnum prunifolium*

Black willow *Salix nigra*

Bloodroot *Sanguinaria canadensis*

Blueberry *Vaccinium* sp.

Blue toadflax; Canada toadflax *Nuttallanthus canadensis*

Blunt broom sedge *Carex tribuloides*

Bottlebrush sedge; bearded sedge *Carex hystericina*

Bouncingbet; soapwort; goodbye summer *Saponaria officinalis*

Broadleaf arrowhead; duck potato *Sagittaria latifolia*

Broadleaf cattail *Typha latifolia*

Broom sedge *Carex scoparia*

Burr marigold; bearded beggarticks *Bidens aristosa*

Buttercup *Ranunculus* sp.

Butterflybush *Buddleja* sp.

Butterfly milkweed *Asclepias tuberosa*

Callery pear *Pyrus calleryana*

Canadian serviceberry; shadbush *Amelanchier canadensis*

Cardinalflower *Lobelia cardinalis*

Carolina elephantsfoot *Elephantopus carolinianus*

Chestnut oak *Quercus montana*

Common boneset; thoroughwort *Eupatorium perfoliatum*

Common buttonbush *Cephalanthus occidentalis*

Common chickweed *Stellaria media*

Common dandelion *Taraxacum officinale*

Common duckweed *Lemna minor*

Common evening primrose *Oenothera biennis*

Common milkweed *Asclepias syriaca*

Common mullein *Verbascum thapsus*

Common pepperweed *Lepidium densiflorum*

Common rush; soft rush *Juncus effusus*

Common sneezeweed *Helenium autumnale*

Common winterberry *Ilex verticillata*

Coneflower *Dracopis* sp.

Coon's tail; hornwort *Ceratophyllum demersum*

Cranefly orchid; crippled cranefly *Tipularia discolor*

Crimsoneyed rosemallow *Hibiscus moscheutos*

Curly pondweed *Potamogeton crispus*

Devil's walkingstick *Aralia spinosa*

Dotted smartweed *Polygonum punctatum*

Drooping sedge *Carex prasina*

Dutchman's breeches *Dicentra cucullaria*

Eastern cottonwood *Populus deltoides*

Eastern redbud *Cercis canadensis*

False Solomon's seal *Maianthemum racemosum*

Flowering dogwood *Cornus florida*

Fringed loosestrife *Lysimachia ciliata*

Garlic mustard *Alliaria petiolata*

Golden ragwort *Packera aurea*

Gray dogwood *Cornus racemosa*

Great ragweed *Ambrosia trifida*

Green arrow arum *Peltandra virginica*

Green ash *Fraxinus pennsylvanica*

Greenbriar *Smilax* sp.

Groundnut *Apios americana*

Halberdleaf rosemallow *Hibiscus laevis*

Halberdleaf tearthumb *Polygonum arifolium*

Hawkweed *Hieracium* sp.

Hawthorn *Crataegus* sp.

Hemlock *Tsuga* sp.

Hepatica; liverleaf *Hepatica nobilis*

Hickory *Carya* sp.

Highbush blueberry *Vaccinium corymbosum*

Indianpipe *Monotropa uniflora*

Jack-in-the-pulpit *Arisaema triphyllum*

Japanese honeysuckle *Lonicera japonica*

Japanese stiltgrass; Nepalese browntop *Microstegium vimineum*

Jewelweed; spotted touch-me-not *Impatiens capensis*

Lantana *Lantana* sp.

Lettuce *Lactuca* sp.

Lizard's tail; water-dragon *Saururus cernuus*

Loblolly pine *Pinus taeda*

Longbranch frostweed *Helianthemum canadense*

Longhair sedge *Carex comosa*

Lyre-leaved rockcress; lyrate rockcress *Arabis lyrata*

Mapleleaf viburnum *Viburnum acerifolium*

Matgrass *Nardus stricta*

Mockernut hickory *Carya tomentosa*

Mountain laurel *Kalmia latifolia*

Mouseear cress; thale cress *Arabidopsis thaliana*

Nakedflower ticktrefoil *Desmodium nudiflorum*

Narrowleaf cattail *Typha angustifolia*

Northern bayberry *Morella pensylvanica*

Northern red oak *Quercus rubra*

Northern spicebush *Lindera benzoin*

Oriental bittersweet *Celastrus orbiculatus*

Partridgeberry *Mitchella repens*

Pawpaw *Asimina triloba*

Phragmites; common reed *Phragmites australis*

Pickerelweed *Pontederia cordata*

Pine *Pinus* sp.

Pink lady's slipper; moccasin flower *Cypripedium acaule*

Pipsissewa *Chimaphila umbellata*

Plantain *Plantago* sp.

Pond apple *Annona glabra*

Pondweed; waterweed *Elodea* sp.

Porcelain berry; amur peppervine *Ampelopsis brevipedunculata*

Prickly sedge *Carex stipata*

Pumpkin ash; red ash *Fraxinus profunda*

Purple chokeberry *Aronia* × *prunifolia*

Purpleleaf willowherb *Epilobium coloratum*

Purple loosestrife *Lythrum salicaria*

Purple passionflower; maypop *Passiflora incarnata*

Purplestem aster; marsh aster *Symphyotrichum puniceum*

Red maple *Acer rubrum*

Redwood *Eucalyptus transcontinentalis*

Reed canarygrass *Phalaris arundinacea*

Rice cutgrass *Leersia oryzoides*

River birch *Betula nigra*

Sandbox tree; dynamite tree *Hura crepitans*

Sassafras *Sassafras albidum*

Scarlet oak *Quercus coccinea*

Sedge *Carex* sp.

Shoreline sedge *Carex hyalinolepis*

Showy orchid *Galearis spectabilis*

Silky dogwood *Cornus amomum*

Skunk cabbage *Symplocarpus foetidus*

Smooth alder; hazel alder *Alnus serrulata*

Smoothsheaf sedge *Carex laevivaginata*

Softstem bulrush *Schoenoplectus tabernaemontani*

Southern red oak *Quercus falcata*

Spatterdock; yellow pond-lily *Nuphar lutea advena*

Sphagnum moss; peat moss *Sphagnum* sp.

Spotted knapweed *Centaurea maculosa*

Spotted wintergreen; striped prince's pine *Chimaphila maculata*

Spreading chervil *Chaerophyllum procumbens*

Staghorn sumac *Rhus typhina*

Star of Bethlehem orchid *Angraecum sesquipedale*

Swamp chestnut oak *Quercus michauxii*

Swamp milkweed *Asclepias incarnata*

Swamp rose *Rosa palustris*

Swamp rosemallow *Hibiscus grandiflorus*

Sweetbay; sweetbay magnolia *Magnolia virginiana*

Sweet cicely; anise *Myrrhis odorata*

Sweetflag *Acorus americanus*

Sweetscented joe pye weed; trumpet weed *Eutrochium purpureum*

Thread-leaf naiad; slender waternymph *Najas gracillima*

Thymeleaf sandwort *Arenaria serpyllifolia*

Tidalmarsh amaranth; waterhemp *Amaranthus cannabinus*

Toothwort *Cardamine* sp.

Tree of heaven *Ailanthus altissima*

Trillium *Trillium* sp.

Tuliptree; tulip poplar *Liriodendron tulipifera*

Violet *Viola* sp.

Virginia creeper *Parthenocissus quinquefolia*

Virginia dayflower; spiderwort *Commelina virginica*

Virginia springbeauty; fairy spud *Claytonia virginica*

Water millet; giant cutgrass *Zizaniopsis miliacea*

Waterpepper; marshpepper knotweed; swamp smartweed *Polygonum hydropiper*

Waterthyme; hydrilla *Hydrilla verticillata*

White fringetree *Chionanthus virginicus*

White oak *Quercus alba*

Wild ginger *Asarum* sp.

Wild lettuce; bitter lettuce *Latuca virose*

Wild rice; wildrice *Zizania* sp.

Wild rye *Elymus* sp.

Wingstem *Verbesina alternifolia*

Wreath goldenrod *Solidago caesia*

Yellow waterlily *Nymphaea mexicana*

Yew *Taxus* sp.

Zinnia *Zinnia* sp.

Notes

THE SETTING

1. Although the Sanctuary comprises a network of critical habitats along the midpoint of the Patuxent River, including the Sanctuary proper, the Glendening Nature Preserve, and Wooton's Landing, the vast majority of my observations were made at the Sanctuary proper.

2. Early colonial settlement is evident today in Old English names. The word "Patuxent," however, is derived from the Algonquin tongue and means "water running over loose stones."

3. Wetlands also play a number of beneficial roles, principally for water purification, flood control, and shoreline stability, as well as acting as a carbon sink that absorbs and stores carbon dioxide from the atmosphere.

JANUARY: THE COLD OR SNOW MOON

1. John Mitchell created the most comprehensive and probably the largest eighteenth-century map of eastern North America. First published in 1755, in conjunction with the imminent Seven Years' War, the map was used during the Treaty of Paris (1783) to define the boundaries of the newly independent United States. To this day, it is still employed to resolve border disputes.

FEBRUARY: THE HUNGER MOON

1. Later, I examine the female catkins of river birches for their tiny "seed-birds." Each catkin contains many small flowers (florets)—bounded by scales of a bright yellow hue, often tinged with red—that become brown and woody in late winter. The minute seeds within are windborne by the wing-like margins on the outer coat of the shells enclosing a flattened nut. I shake some catkins, and a shower of "seed-birds" falls in my hands. I intrude further and pinch off the little scale at a catkin's tip. It frees the entire mass of seeds, and they fall in a steady cascade until only a bare stem is left.

2. This squirrel is one of the few mammalian species able to descend a tree headfirst. It does so by rotating its feet, so the claws of its hind paws point backward and thereby grip the tree bark.

3. Nuts that were experimentally cached, however, were recovered at the same rate as naturally cached nuts, indicating that memory is not as important in locating nuts as has been suggested. Thus squirrels seem to be able to locate caches by olfactory clues, aided by a shallow burial of the objects (McCracken [undated]).

4. Mind thinking refers to an ability to attribute mental states to oneself, and to others, and to understand that others have beliefs, desires, intentions, and perspectives different from one's own.

MARCH: THE WAKENING OR CROW MOON

1. In 2017, ospreys first arrived at Jug Bay on 22 February and reinforce the trend of the last 30 years—nearly every year arriving earlier than the year before. Greg Kearns, longtime naturalist at Patuxent River Park, across the river from Jug Bay, views them as adaptors: they know when it is time to return to their summer haunts and, as our winters become milder, they arrive earlier. Cold snaps do not deter them, as perch and gizzard shad—their major food sources—begin spawning during the warming winter months.

2. One of the many results reported in the Monitoring Avian Productivity and Survivorship (MAPS) study at the Sanctuary recorded a decline in the Carolina wren population following severe winters with much ice.

3. American mud turtles often emerge earlier than any other species at Jug Bay, being built to handle very cold temperatures. They could be given a lift to the Patuxent River by the tidal waters of Chesapeake Bay, where they would stay quite warm in the stable temperatures of the river.

APRIL: THE GRASS MOON

1. If wooden posts are used in place of metal poles for nest boxes, they require a guard to keep out raccoons and other predators, such as snakes.

2. February's pollen peak was the highest in 20 years, including that from species of cedar, cypress, elm, and maple trees and from uncommonly early grasses. Now April's pollen types include those from species of ash, birch, oak, and mulberry trees. Unlike spring ephemerals, summer greens can tolerate shade and assimilate some carbon in the low light conditions under a closed canopy.

3. Woodpeckers experience forces of up to 1200G to 1400G while pecking or drumming. Research workers at the Boston University School of Medicine studied the pickled brains of woodpeckers to see how drumming affects the bird's head. They found that, over time, woodpeckers build up a large amount of a protein called tau (associated with brain damage in humans), but the birds do not appear to suffer adverse effects. One of the members of the study team, Peter Cummings, wondered, "If the tau accumulation is a protective adaptation, is there something we can pick out to help humans with neurodegenerative diseases?" (Farah, Siwek, and Cummings 2018).

4. Alexander Pope's verse laid the foundation for one of the most widely agreed-on principles of landscape architecture: landscape designs should always be adapted to the context in which they are located.

5. Studies have found that *Bacillus licheniformis* can cause feather degradation in birds (e.g., Burtt and Ichida 1999).

6. According to a *Scientific American* blog, "the Environmental Protection Agency estimates that gas-powered lawnmowers—which emit 11 times more air pollution than a

new car for every hour of operation—contribute as much as 5 percent of the smog in some areas of the United States. Every summer, Americans spill 17,000,000 gallons of gasoline when refueling mowers and other garden equipment. Moreover, of the approximately 90 million American households with a yard or garden, 45 million use chemical fertilizers, 46 million use insecticides, and 47 million use chemical weedkillers. Such chemicals—many of which, especially older varieties, have known health risks—contaminate natural habitats and seep into our homes and drinking water" (Jabr 2013).

MAY: THE PLANTING MOON

1. A study on the effects of noise from natural gas fields in New Mexico has shown that western bluebirds find it harder to hear the sounds of approaching predators, or even normal noises of the surrounding world. Consequently, these birds have to maintain constant vigilance, and the level of corticosterone (a stress-related hormone) in adults caused their health to deteriorate, along with that of their nestlings (Kleist et al. 2018).

2. As the weeks progressed, migration proved to be bizarre and unpredictable. Though southwesterly winds blew in April and migrations that month followed the norm, cold, strong, northwesterly winds blew for 15 days in May, with birds being stalled or pushed around. Weather fronts ran from east to west, blocking the birds' normal passage. In a few places, migrants arrived early, some appeared in unexpected concentrations, and others came in low numbers. More vagrants showed up, such as fork-tailed flycatchers. Birds flew at low altitudes to counter the adverse conditions, with significantly more of them crashing into buildings and dying. Such blockage was eventually uncorked, and a normal, if late, migration pattern resumed.

3. Only four species—least bitterns, marsh wrens, common yellowthroats, and red-winged blackbirds—breed in the vegetation of Jug Bay's tidal wetlands. One of the reasons is that few species can build nests above a tidal amplitude of nearly three feet and bouts of rain during a full-moon tide.

JUNE: THE ROSE MOON

1. Male eastern tiger swallowtails participate in a behavior called puddling, in which they congregate on mud or puddles. They extract salts and amino acids from these sources and transfer them during mating, to help provide nutrients for the eggs.

JULY: THE HEAT MOON

1. The name "rosemallow" has its origin in Europe, where people would collect these plants from marshy areas. The roots would then be boiled to produce a white sticky substance. Though not thought to provide medical benefits, it still satisfied the sweet tooth of children, hence "mallow." One Native American tribe used this plant to cure inflamed bladders.

2. The unique appearance of the flowers of purple passionflowers was purported, by early Spanish explorers, to represent the sufferings of Christ.

3. Sad to note, when I returned a few days later, the nest was in tatters on the floor of the blind. It was my hope that the barn swallows flying within my sight included some of these young.

4. Tree swallows are farsighted in one eye and nearsighted in the other, enabling them to spot insects at a distance and yet capture them at point blank range.

5. The Maryland Conservation Corps is an award-winning AmeriCorps program that engages young adults in extensive natural resource management and park conservation projects.

6. Gypsy moth habitat overlaps with that of northern tiger swallowtails, and experiments indicate that a gypsy moth's pathogens and bodily fluids negatively affect the survival of swallowtail larvae.

7. Damselflies and dragonflies are hemimetabolous, that is, insects having incomplete metamorphosis, with no pupal stage in their development.

8. Diving spiders use a somewhat similar mechanism, although they also undertake something quite unique. They build a special silk web underwater, fill it with air carried from the surface, and use the air stored in this silk balloon (the diving bell) to breathe. Nonetheless, neither the air film carried by the spiders on their bodies nor the air contained within the diving bell is enough to completely satisfy their oxygen requirements. Consequently, they depend on occasional surface trips to survive, with bristles that fold to aid their swimming action.

August: The Thunder Moon

1. In addition, gallic acid, released by phragmites, is degraded by ultraviolet light, thus producing mesooxalic acid, effectively hitting susceptible plants and seedlings with two harmful toxins.

2. In contrast, the buoyant seeds of green arrow arums float in and out with the tide and thereby disperse over a wide area.

3. Wood infected with the actively growing mycelium of honey mushrooms is luminous when seen in total darkness. In the United Kingdom during World War II, firewatchers in lumberyards observed this luminous wood on several occasions and attributed the glow to the presence of phosphorus from incendiary bombs.

4. The doctrine of signatures is the ancient notion proposing that a Creator provided clues about which plants may be used to treat particular illnesses by shaping the plants to resemble the organs or conditions they should be used for. A few plants thus identified turned out by chance to have some value in treating the similar-appearing body part or condition. It is likely that many valuable herbs were in use before the doctrine and that the organ-to-plant match was made later to accommodate and validate the doctrine. While the

doctrine of signatures has been totally debunked, many herbal remedies employed today owe their use to this notion.

5. Note that in older English usage, "thorough" was not distinguished from "through."

SEPTEMBER: THE HUNTING MOON

1. "Knap" most likely refers to the fuzzy texture of carpets when their knot loops are cut.

2. Three or more branched flower clusters on each wild rice plant bear an average of 600 dangling seeds.

3. "Dog days" refers to the hot, sultry days of summer. Historically, they referred to the time period when the star Sirius (the "dog star") rises and sets in conjunction with the sun. Greek and Roman astrologers connected this with heat, drought, sudden thunderstorms, lethargy, fever, mad dogs, and bad luck.

4. North American beavers may kill conifers by girdling them (removing a strip of bark, including its inner layers, around the entire circumference of the tree), without either eating the bark or attempting to cut the tree down, thus reducing competition in favor of their preferred tree species.

5. Net primary production is the difference between the rate at which all plants in an ecosystem produce useful chemical energy and the rate at which they use some of that energy during respiration.

6. A photograph of a monarch butterfly by Marylander Karen Maynard is featured in the US Postal Service's new "protect pollination" series of forever stamps.

7. Pupae are the most common stage of midges for fish to eat, though the adults may be consumed at times.

8. The term "thin as a rail" applies to soras as well as other species of rails.

OCTOBER: THE LEAF-FALLING MOON

1. The Cambridge Conservation Initiative is a unique collaboration between the University of Cambridge and leading internationally focused biodiversity conservation organizations clustered in and around Cambridge in the United Kingdom.

2. Plants that produce a heavy set of fruits tend to have lighter-colored foliage, largely due to the availability of nitrogen and its movement within the plant. If its roots cannot obtain sufficient quantities of nitrogen in the soil in order to develop fruit and seeds, a plant will drain nitrogen from its leaves. This has a significant impact on chlorophyll content. For example, the dark green of American holly leaves contain four percent nitrogen; with smaller percentages, its older leaves are the first to turn yellow green.

3. Gall midge larvae feeding on plants frequently cause aborted or damaged fruits.

4. Water quality anywhere above the 90 percent oxygen saturation range is considered to be satisfactory for living organisms.

5. Unlike sap drawn from the xylem (sapwood) of a tree, which is tapped in the deeper

holes drilled to make maple syrup and contains from 2 to 3 percent sugar, phloem sap may consist of up to 20–30 percent sugar.

6. Blonder et al. (2011) contend that a good understanding of leaves will become incorporated into climate models. This not only can help balance the carbon budget, but also predict evaporation rates and other weather- and climate-related matters that are heavily reliant on plants.

7. The smoothness of beech bark invites carving, and the very word "beech" comes from the Anglo-Saxon *boc*, meaning "word" or "letter," from which "book" is derived. Before papyrus, beech bark was harvested as sheets for use in writing tablets. A stylus was used to etch straight lines, following the grain. Circles had to be formed as a series of corners. Consequently, in the Middle Ages, the Gothic style of writing evolved from letters composed of angles.

8. A study in Virginia documented 50 blue jays caching 150,000 acorns a month—an average of about 110 acorns per bird per day (Darley-Hill and Johnson 1981). Blue jays make choices about which trees the acorns are harvested from, and which and how many nuts are gathered. Nearly all acorns that were not harvested and cached were either consumed immediately by the blue jays or destroyed by weevils.

NOVEMBER: THE BEAVER MOON

1. Marbled salamanders breed in autumn—a reproductive strategy very different from other related species, such as spotted salamanders, which generally breed in the spring.

2. English folklore holds that Jack Frost, an elfish creature, is responsible for the feathery patterns of frost found on windows on cold mornings.

3. Why the name "witchhazel"? Most likely it is because the plant was used to make divining rods to find witches and water, and even gold.

4. A debate prevails over abolishing the hunting of tundra swans, because few sportsmen can distinguish them in flight from at-risk trumpeter swans.

5. The colors green, orange, and red appear to a deer as shades of gray, which is why hunters can wear bright orange safety clothing.

DECEMBER: THE LONG NIGHT MOON

1. One study examined reports of traffic accidents involving four species of ungulates, including white-tailed deer in New York State, over a two-year period and found that collisions were most frequent during the full phase of the lunar cycle (Colino-Rabanal et al. 2017). The pattern is evident throughout the year but is stronger during some months, occurring during full moon nights. Deer matings, however, are greatest around the new moon. In addition, insect larvae on the water's surface transform into winged adults just before sunrise during a two-week period in late May or early June, having to perform their mating flight, copulate, and lay eggs within one to two hours (Sweeney and Vannote 1982).

2. When certain plants, such as lettuces, grow too quickly, the stem lengthens, a condition known as bolting.

3. Graupel is snow coated in supercooled liquid, typically forming as precipitation when the lower atmosphere is unstable.

4. The brown, cigar-shaped flower heads of cattails contain up to 125,000 seeds per head.

5. Beavers' caches (often thin branches of tasty shrubs, such as willows) are placed underwater, anchored in mud at the bottom of the pond near the entrance to the lodge. These will be used in winter, but beavers will continue to cut fresh logs when conditions are free of ice.

6. Castoreum is the exudate from castor sacs, used as a tincture (an alcohol-based extract) in some perfumes and as a food additive.

EPILOGUE

1. The National Aeronautics and Space Administration (NASA) reported that Earth's global surface temperatures in 2017 ranked as either the second warmest (NASA's analysis) or third warmest (analysis by the National Oceanic and Atmospheric Administration) recorded since measurements began being taken 138 years ago.

Bibliography

GENERAL READING

Note: Readers may wish to consult *Marsh Notes*, produced quarterly by the Friends of Jug Bay, and the research reports produced by Sanctuary staff and volunteers. Both are available on the Jug Bay Wetlands Sanctuary website at www.jugbay.org.

Aleksuik, M. 1968. "Scent-mound communication, territoriality and population regulation in beaver (*Castor canadensus* Kuhl)." *Journal of Mammalogy* 49: 759–762.

Alexander, R. M. 2003. *Principles of Animal Locomotion*. Princeton University Press, Princeton, NJ.

American Society for Microbiology. 2004. "Birds use herbs to protect their nests." Science Daily, 27 May, https://www.sciencedaily.com/releases/2004/05/040527080935.htm.

Andersson, J., Borg-Karlson, A.-K., and Wiklund, C. 2000. "Sexual cooperation and conflict in butterflies: A male-transferred anti-aphrodisiac reduces harassment of recently mated females." *Proceedings of the Royal Society B: Biological Sciences* 267 (1450): 1271–1275. doi:10.1098/rspb.2000.1138. Also available online at PubMed, PMC 1690675, PMID 10972120.

Angier, N. 1994. "To remember seed caches, bird grows new brain cells." *New York Times*, 15 November, https://www.nytimes.com/1994/11/15/science/to-remember-seed-caches-bird-grows-new-brain-cells.html.

Appel, H. M., and Cocroft, R. B. 2014. "Plants respond to leaf vibrations caused by insectivore chewing." *Oecologia* 175 (4): 1257–1266. doi:10.1007/s00442-014-2995-6.

Araya-Salas, M. 2012. "Is birdsong music? Evaluating harmonic intervals in songs of a neotropical songbird." *Animal Behaviour* 84 (2): 309–313.

Audubon, J. J. [multiple dates]. *Birds of America*. Available in many editions. His color plates and text can also be viewed at the National Audubon Society's website, https://www.audubon.org/birds-of-america/.

Barendregt, A., Whigham, D. F., and Baldwin, A. H. 2009. *Tidal Freshwater Wetlands*. Backhuys, Leiden, Netherlands.

Bear, I. J., and Thomas, R. G. 1964. "Nature of argillaceous odour." *Nature* 201 (4923): 993–995. doi:10.1038/201993a0.

Beehler, B. M. 2018. *North on the Wing: Travels with the Songbird Migration of Spring*. Smithsonian Books, Washington, DC.

Bent, A. C. 1948. *Life Histories of North American Nuthatches, Wrens, Thrashers and Their Allies: Order Passeriformes*. US Government Printing Office, Washington, DC.

Blonder, B., Violle, C., Bentley, L. P., and Enquist, B. J. 2011. "Venation networks and the origin of the leaf economics spectrum." *Ecology Letters* 14 (2): 91–100.

Borenstein, S. 2017a. "Study: Climate change goosed odds of freakishly hot February." Associated Press, 18 January, https://apnews.com/0cc9e34e5172443881 56a134e9200 566/.

————. 2017b. "Science says: Jack Frost nipping at your nose ever later." Associated Press, 27 October, https://apnews.com/664a16182cbc4e3e9a26957f3cddc927.

Borland, H. 1964. *Sundial of the Seasons*. Lippincott, Philadelphia, PA.

Bragg, D. C., Shelton, M. G., and Zeide, B. 2003. "Impacts and management implications of ice storms on forests in the southern United States." *Forest Ecology and Management* 186: 99–123.

Breitburg, D., Jordan, T. E., and Lipton, D. 2003. "Preface." In "From Ecology to Economics: Tracing Human Influence in the Patuxent River Estuary and its Watershed," dedicated issue. *Estuaries* 26 (2A): 167–170.

Brinson, M. M., Swift, B. L., Plantico, R. C., and Barclay, J. S. 1981. *Riparian Ecosystems: Their Ecology and Status*. Biological Services Program FWS/OBS-81/17. US Fish and Wildlife Service, Washington, DC.

Brinson, M. M., and Verhoeven, J. 1999. "Riparian forests." Pp. 265–299 in *Maintaining Biodiversity in Forest Ecosystems*, ed. M. Hunter Jr. Cambridge University Press, New York, NY.

Burroughs, J. 1886. *Signs and Seasons*. Houghton Mifflin, Boston, MA.

Burtt, E. H., and Ichida, J. M. 1999. "Occurrence of feather-degrading bacilli in the plumage of birds." *Auk* 116 (2): 364–372.

Caduto, M. J. 2016. *Through a Naturalist's Eyes: Exploring the Nature of New England*. University Press of New England, Lebanon, NH.

Capra, F. 1996. *The Web of Life: A New Synthesis of Mind and Matter*, 1st Anchor Books edition. Doubleday, New York, NY.

Carroll, D. M. 1991. *The Year of the Turtle: A Natural History*. Camden House, Charlotte, VT.

————.2001. *Swampwalker's Journal: A Wetlands Year*, 1st Mariner Books edition. Houghton Mifflin, Boston, MA.

Carson, R. 1962. *Silent Spring*. Houghton Mifflin, New York, NY.

————. 1965. *A Sense of Wonder*. Harper & Row, New York, NY.

Ceballos, G., Ehrlich, P. R., and Dirzo, R. 2017. "Biological annihilation via the ongoing sixth mass extinction signaled by vertebrate population losses and declines." *Proceedings of the National Academy of Sciences* 224 (30): E6089–E6096.

Climate Central. 2015. "Across U.S., heaviest downpours on the rise." Climate Central, 27 May, www.climatecentral.org/news/across-us-heaviest-downpours-on-the-rise -18989/.

Colbeck, S. C. 1986. "Classification of seasonal snow cover crystals." *Water Resources Research* 22: 59S–70S.

Colino-Rabanal, V. J., Langen, T. A., Peris, S. J., et al. 2017. "Ungulate-vehicle collision rates associated with the phase of the moon." *Biodiversity and Conservation* 24 (3): 681–694. https://doi.org/10.1007/s10531-017-1458-x/.

Cooper-White, M. 2015. "Here's why rain has that distinctive smell." *Huffpost*, 17 January, https://www.huffingtonpost.com/2015/01/15/why-smell-rain-high-speed-video_n_6479666.html.

Corfidi, S. F. 2014. "The colors of sunset and twilight." NOAA/NWS Storm Prediction Center, https://www.spc.noaa.gov/publications/corfidi/sunset/.

Costanzo, J. P., Lee, R. E., Jr., and Lortz, P. H. 1993. "Glucose concentration regulates freeze tolerance in the wood frog *Rana sylvatica*." *Journal of Experimental Biology* 181: 245–255.

Cox, D. D. 2002. *A Naturalist's Guide to Wetland Plants*. Syracuse University Press, Syracuse, NY.

Crumley, J. 2017. *The Nature of Winter*. Saraband, Glasgow, UK.

Curtain, P. D., Brush, G. S., and Fisher, G. W. 2001. *Discovering the Chesapeake Bay: The History of an Ecosystem*. Johns Hopkins University Press, Baltimore, MD.

Darley-Hill, S., and Johnson, W. C. 1981. "Acorn dispersal by the blue jay (*Cyanocitta cristata*)." *Oecologia* 50 (2): 231–232.

Darwin, C. 1859. *On the Origin of Species by Means of Natural Selection; or, The Preservation of Favoured Races in the Struggle for Life*. John Murray, London, UK.

———. 1862. *On the Various Contrivances by which British and Foreign Orchids Are Fertilised by Insects, and on the Good Effects of Intercrossing*. John Murray, London, UK.

———. 1871. *On the Descent of Man, and Selections in Relation to Sex*. John Murray, London, UK.

Das, T. M. 1979. "What is the value of a tree?" *Indian Biologist* 11 (1–2): 73–79.

Dawson, W. R., and Carey, C. 1976. "Seasonal acclimatization to temperature in carduline finches: I, Insulative and metabolic adjustments." *Journal of Comparative Physiology B: Biochemical, Systemic, and Environmental Physiology* 112: 317–333.

Delgado, P., ed. 2011. *Chesapeake Bay National Estuarine Research Reserve in Maryland: A Site Profile*. Maryland Department of Natural Resources; Chesapeake and Coastal Service, Coastal Zone Management; and Chesapeake Bay National Estuarine Research Reserve, Annapolis, MD. http://dnr.maryland.gov/waters/cbnerr/Documents/publications/CBNERRMD_SiteProfile_Dec2011.pdf.

Doolittle, E., and Brumm, H. 2012. "O canto do uirapuru: Consonant intervals and patterns in the song of the musician wren." *Journal of Interdisciplinary Music Studies* 6 (1): 55–85.

Eastman, J. 1995. *The Book of Swamp and Bog: Trees, Shrubs, and Wildflowers of Eastern Freshwater Wetlands*. Stackpole Books, Mechanicsburg, PA.

Ehrlich, P. R., Dobkin, D. S., and Wheye, D. 1988. "Mixed-species flocking." https://web.stanford.edu/group/stanfordbirds/text/essays/Mixed-Species_Flocking.html.

Einstein, A. 1905. "On the motion of small particles suspended in liquids at rest required by the molecular-kinetic theory of heat." *Annalen der Physik* 17: 549–560.

Elliot, L., and Hershberger, W. 2007. *The Songs of Insects*. Houghton Mifflin, Boston, MA.

Ellison, W. G., ed. 2010. *Second Atlas of the Breeding Birds of Maryland and the District of Columbia*. Johns Hopkins University Press, Baltimore, MD.

Ernst, C. H. 1976. "Ecology of the spotted turtle, *Clemmys guttata* (Reptilia, Testudines, Testudinidae), in southeastern Pennsylvania." *Journal of Herpetology* 10 (1): 25–33. doi:10.2307/1562924.

Faccio, S. D. 2003. "Effects of ice storm–created gaps on forest breeding bird communities in central Vermont." *Forest Ecology and Management* 186 (1–3): 133–145.

Farah, G., Siwek, D., and Cummings, P. 2018. "Tau accumulations in the brains of woodpeckers." *PLoS ONE* 13 (2): e0191526. https://doi.org/10.1371/journal.pone.0191526/.

Farmer, E. E. 2014. *Leaf Defence*. Oxford University Press, New York, NY.

Farmer, E. E., and Mueller, M. J. 2013. "ROS-mediated lipid peroxidation and RES-activated signaling." *Annual Review of Plant Biology* 64 (1): 429–450.

Formozov, A. N. 1969. *Snow Cover as an Integral Factor of the Environment and Its Importance in the Ecology of Mammals and Birds*. Boreal Institute, University of Alberta, Edmonton, AB.

Foden, S. [undated]. "Why do a dove's wings whistle?" Pets, http://animals.mom.me /doves-wings-whistle-8649.html.

Fox, J. F. 1982. "Adaptation of gray squirrel behavior to autumn germination by white oak acorns." *Evolution* 36 (4): 800–809.

Fritz, A. 2017. "Where's winter? Here's why this season has been so 'strange' on the East Coast." *Washington Post*, 6 February.

Gibbons, E. 1962. *Stalking the Wild Asparagus*. Alan C. Hood, Chambersburg, PA.

Gillium, F. S. 2014. *The Herbaceous Layer in Forests of Eastern North America*. Oxford University Press, New York, NY.

Gooley, T. 2016. *How to Read Water: Clues and Patterns from Puddles to the Sea*. Sceptre, London, UK.

Grant, S. 2004. "The squirrel's bag of tricks: They can't get out of the way of cars, but other behaviors demonstrate advanced thinking (for a rodent)." *Harford Courant*, 21 October.

Grunert, J. [undated]. "What attracts male moths to female moths?" Pets, http://animals .mom.me/attracts-male-moths-female-moths-7524.html.

Halfpenny, J. C., and Ozanne, R. D. 1989. *Winter: An Ecological Handbook*. Johnson Books, Boulder, CO.

Hall, C. C. 1910. "A Relation of Maryland, 1635." Pp. 63–112 in *Narratives of Early Maryland, 1633–1634*, ed. C. C. Hall. Barnes & Noble, New York, NY.

Hamblyn, R. 2001. *The Invention of Clouds: How an Amateur Meteorologist Forged the Language of the Skies*. Farrar, Straus & Giroux, New York, NY.

Hartshorne, C. 1992. *Born to Sing: An Interpretation and World Survey of Bird Song*. Indiana University Press, Bloomington, IN.

Haskell, D. G. 2012. *The Forest Unseen: A Year's Watch in Nature.* Viking, New York, NY.

Heinrich, B. 1997. *The Trees in My Forest.* HarperCollins, New York, NY.

———. 2003. *Winter World: The Ingenuity of Animal Survival.* HarperCollins, New York, NY.

———. 2009. *Summer World: A Season of Bounty.* HarperCollins, New York, NY.

———. 2010. *The Nesting Season: Cuckoos, Cuckolds, and the Invention of Monogamy.* Belknap Press of Harvard University Press, Cambridge, MA.

———. 2014. *The Homing Instinct: Meaning and Mystery in Animal Migration*, 1st Mariner Books edition. Houghton Mifflin Harcourt, Boston, MA.

Helmenstine, A. M. 2017. "How much oxygen does one tree produce?" ThoughtCo, 10 April, https://www.thoughtco.com/how-much-oxygen-does-one-tree-produce-606785/.

Henderson, C. 2017. *A New Map of Wonders.* University of Chicago Press, Chicago, IL.

Higgins, R. 2017. *Thoreau and the Language of Trees.* University of California Press, Oakland, CA.

Hill, E. P. 1982. "Beaver." Pp. 256–281 in *Wild Mammals of North America: Biology, Management, and Conservation*, ed. J. A. Chapman and G. A. Feldhamer. Johns Hopkins University Press, Baltimore, MD.

Hingee, M., and Magrath, R. D. "Flights of fear: A mechanical wing whistle sounds the alarm in a flocking bird." *Proceedings of the Royal Society B: Biological Sciences* 276 (1676): 4173–4179.

Horton, K. G., Van Doren, B. M., La Sorte, F. A., et al. 2018. "Navigating north: How body mass and winds shape avian flight behaviours across a North American migratory flyway." *Ecology Letters* 21 (7): 1055–1064. doi:10.1111/ele.12971.

Howard, L. 1803. *On the Modifications of Clouds: And on the Principles of Their Production, Suspension, and Destruction; Being the Substance of an Essay Read before the Askesian Society in the Session 1802–3.* J. Taylor, London, UK.

Hynes, H. B. N. 1970. *The Ecology of Running Waters.* University of Toronto Press, Toronto, ON.

Jabr, F. 2013. "Outgrowing the traditional grass lawn." Brainwaves, *Scientific American Blog*, 29 July, https://blogs.scientificamerican.com/brainwaves/outgrowing-the-traditional-grass-lawn/.

Jensen, K. H. 2016. "Pollen, water, and wind: Chaotic mixing in a puddle of water." *Physical Review Fluids* 1 (5), article 050507.

Jirinec, V., Cristo, D. A., and Leu, M. 2017. "Songbird community varies with deer use in a fragmented landscape." *Landscape and Urban Planning* 161: 1–9.

Johns Hopkins University. 1966. *Report on the Patuxent River Basin, Maryland.* Prepared by the members of the Water Management Seminar, The Johns Hopkins University.

Johnson, C., and Adkisson, C. 1986. "Airlifting the oaks." *Natural History* 10 (86): 41–46.

Johnson, G. 1926. *Nature's Program.* Nelson Doubleday, Garden City, NY.

Kamisugi, Y., Schaefer, D. G., Kozak, J., et al. 2012. "MRE11 and RAD50, but not NBS1, are essential for gene targeting in the moss *Physcomitrella patens.*" *Nucleic Acids Research* 40 (8): 3496–3510. doi:10.1093/nar/gkr1272. Also available online at PubMed, PMC 3333855, PMID 22210882.

Karageorgou, P., and Manetas, Y. 2006. "The importance of being red when young: Anthrocyanins and the protection of young leaves of *Quercus coccifera* from insect herbivory and excess light." *Tree Physiology* 26: 613–631.

Kenny, L. P., and Burne, M. R. 2009. *A Field Guide to the Animals of Vernal Pools.* Massachusetts Division of Fisheries and Wildlife, Natural Heritage and Endangered Species Program; and Vernal Pool Association, Westborough, MA.

Khan, H., and Brush, G. S. 1994. "Nutrient and metal accumulation in a freshwater tidal marsh." *Estuaries* 17 (2): 345–360.

Kita, M., Nakamura, Y., Ohdachi, S. D., et al. 2004. "*Blarina* toxin, a mammalian lethal venom from the short-tailed shrew *Blarina brevicauda*: Isolation and characterization." *Proceedings of the National Academy of Sciences* 101 (20): 7542–7547. doi:10.1073 /pnas.0402517101. Also available online at PubMed, PMC 419642, PMID 15136743.

Kleist, N. J., Guralnick, R. P., Cruz, A., et al. 2018. "Chronic anthropogenic noise disrupts glucocorticoid signaling and has multiple effects on fitness in an avian community." *Proceedings of the National Academy of Sciences* 115 (4): E648–E657. doi:10.1073 /pnas.1709200115.

Kolbert, E. 2017. "The fate of the Earth." *New Yorker*, 12 October, https://www.newyorker .com/tech/elements/the-fate-of-earth/.

Kress, W. J., and Stine, J. K. 2017. *Living in the Anthropocene: Earth in the Age of Humans.* Smithsonian Books, Washington, DC.

Kronfeld-Schor, N., Dominoni, D., de la Iglesia, H., et al. 2013. "Chronobiology by moonlight." *Proceedings of the Royal Society B: Biological Sciences* 280 (176): 20123088. http: //dx.doi.org/10.1098/rspb.2012.3088/.

Layne, J. R., Jr., and Lee, R. E., Jr. 1995. "Adaptations of frogs to survive freezing." *Climate Research* 5: 53–59.

Leopold, A. 1948. *A Sand County Almanac and Sketches Here and There.* Oxford University Press, New York, NY.

Louv, R. 2005. *Last Child in the Woods: Saving Our Children from Nature Deficit Disorder.* Algonquin Books, Chapel Hill, NC.

———. 2011. *The Nature Principle: Reconnecting with Life in a Digital Age.* Algonquin Books, Chapel Hill, NC.

Lukas, D. 2012. "The cuckoo wasp: A gorgeous parasite." *Bay Nature Magazine*, July–September 2012.

Lund, N. 2014. "The biggest misconception about birds." "Wild Things," *Slate's Animal Blog*, 23 January, www.slate.com/blogs/wild_things/2014/01/23/where_do_birds _sleep_roosting_in_nests_water_flocks_cavities.html.

MacArthur, R. H. 1958. "Population ecology of some warblers of northeastern coniferous forests." *Ecology* 39 (4): 599–619.

MacKay, B. 2013. *A Year across Maryland: A Week-by-Week Guide to Discovering Nature in the Chesapeake Region.* Johns Hopkins University Press, Baltimore, MD.

Mahabale, T. S. 1968. "Spores and pollen grains of water plants and their dispersal." *Review of Palaeobotany and Palynology* 7 (4): 285–296.

Marchand, P. J. 1987. *Life in the Cold: An Introduction to Winter Ecology.* University Press of New England, Hanover, NH.

Markmann-Mulisch, U., Wendeler, E., Zobell, O., et al. 2007. "Differential requirements for RAD51 in *Physcomitrella patens* and *Arabidopsis thaliana* development and DNA damage repair." *Plant Cell* 19 (10): 3080–3089. doi:10.1105/tpc.107.054049. Also available online at PubMed, PMC 2174717, PMID 17921313.

Martell, M. S., Bierregaard, R. O., Washburn, B. E., et al. 2014. "The spring migration of adult North American ospreys." *Journal of Raptor Research* 48 (4): 309–324.

Martin, I. G. 1981. "Venom in the short-tailed shrew (*Blarina brevicauda*) as an insect immobilizing agent." *Journal of Mammology* 62 (1): 189–92.

Maryland Department of Natural Resources. 2016. *Maryland State Wildlife Action Plan 2015–2025.* http://dnr.maryland.gov/wildlife/Pages/plants_wildlife/SWAP_Submission.aspx.

Maryland Department of the Environment, Wetlands and Waterways Program. 2006. *Prioritizing Sites for Wetland Restoration, Mitigation, and Preservation in Maryland,* 31 May. www.mde.state.md.us/programs/Water/WetlandsandWaterways/AboutWetlands/Documents/www.mde.state.md.us/assets/document/wetlandswaterways/AA.pdf.

Mascetti, G. G. 2016. "Unihemispheric sleep and asymmetrical sleep: Behavioral, neurophysiological, and functional perspectives." *Nature and Science of Sleep* 2016 (1): 221–258. Also available online at PubMed, PMC 4948738, PMID 27471418.

Mathews, F. S. 1904. *A Field Book of Wild Birds and Their Music: A Description of the Character and Music of Birds, Intended to Assist In the Identification of Species Common to the United States East of the Rocky Mountains.* Putnam's Sons, New York, NY.

Mayor, S. J., Guralnick, R. P., Tingley, M. W., et al. 2017. "Increasing phonological asynchrony between spring green-up and arrival of migratory birds." *Scientific Reports* 7, article 1902. doi:10.1038/s41598-017-02045-z.

McCracken, B. [undated]. "Do squirrels really know where they bury their food?" Pets, http://animals.mom.me/squirrels-really-bury-food-11108.html.

Meanley, B. 1975. *Birds and Marshes of the Chesapeake Bay Country.* Tidewater, Centreville, MD.

Melillo, J. M., McGuire, A. D., Kicklighter, D. W., et al. 1993. "Global climate change and terrestrial net primary production." *Nature* 363 (6426): 234–240.

Melquist, W. E., Whitman, J. S., and Hornocker, M. C. 1981. "Resource partitioning and coexistence of sympatric mink and river otter populations." Pp. 187–220 in *Proceedings of the Worldwide Furbearer Conference*, ed. J. A. Chapman and D. Pursley. Worldwide Furbearer Conference, Frostburg, MD.

Merlin, C., Gegear, R. J., and Reppert, S. M. 2009. "Antennal circadian clocks coordinate sun compass orientation in migratory monarch butterflies." *Science* 325 (5948): 1700–1704.

Merritt, J. F. 1986. "Winter survival adaptations of the short-tailed shrew (*Blarina brevicauda*) in an Appalachian montane forest." *Journal of Mammalogy* 67 (3): 450–464. doi:10.2307/1381276.

Moyroud, E., Reed, A., Mellers, G., et al. 2017. "Disorder in convergent floral nanostructures enhances signalling to bees." *Nature* 550 (7677): 469–474.

Müller-Schwarze, D. 2009. "Squirrels, acorns, and tannins." Pp. 31–35 in *Hands-On Chemical Ecology*. Springer, New York, NY. https://doi.org/10.1007/978-1-4419-0378-5_6/.

Murdy, E. O. 1997. *Fishes of the Chesapeake Bay*. Smithsonian Institution Press, Washington, DC.

Murray, R. C. 2008. *Tree Biology Notebook: An Introduction to the Science and Ecology of Trees*. STL, Silver Spring, MD.

Murray, T. G., Zeil, J., and Magrath, R. D. 2017. "Sounds of modified flight feathers reliably signal danger in a pigeon." *Current Biology* 27 (22): 3520–3525. http://dx.doi.org/10.1016/j.cub.2017.09.068/.

National Aeronautics and Space Administration (NASA). 2018. "Long-term warming trend continued in 2017: NASA, NOAA." NASA News & Feature Releases, 18 January, https://www.giss.nasa.gov/research/news/20180118/.

National Audubon Society. [undated]. "Climate threatened wood duck." Climate Report, climate.audubon.org/birds/wooduc/wood-duck/.

National Wildlife Federation. [undated]. "Trees and snags." "Cover," National Wildlife Federation, https://www.nwf.org/Garden-for-Wildlife/Cover/Trees-and-Snags.

Nemiller, M. L., Porter, M. L., Keany, J., et al. 2017. "Evaluation of eDNA for groundwater invertebrate detection and monitoring: A case study with endangered *Stygobromus* (Amphipoda: Crangonyctidae)." *Conservation Genetics Resources* 10 (2): 1–11.

Nutt, A. E. 2017. "From cloud nine to climate change, here's why you should always look up." *Washington Post*, 7 October.

Owen, W. 1969. *A Study of the Physical Hydrography of the Patuxent River and its Estuary*. Johns Hopkins University Technical Report 53. Chesapeake Bay Institute, The Johns Hopkins University, Baltimore, MD.

Palmer, J. M., Drees, K. P., Foster, J. T., and Linder, D. L. 2018. "Extreme sensitivity to ultraviolet light in the fungal pathogen causing white-nose syndrome of bats." *Nature Communications* 9, article 35. doi:10.1038/s41467-017-02441-z.

Paradiso, J. L. 1969. *Mammals of Maryland*. North American Fauna No. 66. US Fish and Wildlife Service, Washington, DC.

Pauli, J. N., Zuckerberg, B., Whiteman, J. P., and Porter, W. "The subnivium: A deteriorating seasonal refugium." *Frontiers in Ecology and the Environment* 11 (5): 260–267.

Peattie, D. C. 1950. *A Natural History of Trees of Eastern and Central North America*. Houghton Mifflin, Boston, MA.

Pelton, T. 2018. *The Chesapeake in Focus: Transforming the Natural World*. Johns Hopkins University Press, Baltimore, MD.

Perry, H. R., Jr. 1982. "Muskrats." Pp. 282–325 in *Wild Mammals of North America: Biology, Management, and Conservation*, ed. J. A. Chapman and G. A. Feldhamer. Johns Hopkins University Press, Baltimore, MD.

Peterson, R. T. 1934. *A Field Guide to the Birds*. Houghton Mifflin, Boston, MA.

Purdue University. 2017. "Purdue scientists want help recording wildlife during eclipse." Purdue University Agricultural News, 17 August, https://www.purdue.edu/newsroom/releases/2017/Q3/purdue-scientists-want-help-recording-wildlife-during-eclipse.html.

Reese, J. G. 2017. "Brown creeper (*Certhia americana*) fatality exhibits signs associated with lead poisoning." *Maryland Birdlife* 66 (2): 20–28.

Rensing, S. A., Lang, D., Zimmer, A. D., et al. 2008. "The *Physcomitrella* genome reveals evolutionary insights into the conquest of land by plants." *Science* 319 (5859): 64–69. doi:10.1126/science.1150646. Also available online at PubMed, PMID 18079367.

Robbins, C. S. 1966. *Birds of North America: A Guide to Field Identification*. Golden Press, New York, NY.

———, ed. 1997. *Atlas of the Breeding Birds of Maryland and the District of Columbia*. University of Pittsburg Press, Pittsburgh, PA.

Rock, B. N., ed. 2001. *Preparing for a Changing Climate: The Potential Consequences of Climate Variability and Change; New England Regional Overview*. US Global Change Research Program, Durham, NH.

Rooney, T. P., and Waller, D. M. 2003. "Direct and indirect effects of white-tailed deer in forest ecosystems." *Forest Ecology and Management* 181: 165–176.

Ruskin, J. 1918. *Selections and Essays*. C. Scribner's Sons, New York, NY.

Ruzicka, K. J., Groninger, J. W., and. Zaczek, J. J. 2010. "Deer browsing, forest edge effects, and vegetation dynamics following bottomland forest restoration." *Restoration Ecology* 18 (5): 702–710.

Saltonstall, K., Peterson, P. M., and Soreng, R. J. 2004. "Recognition of *Phragmites australis* subsp. *americanus* (Poaceae: Arundinoideae) in North America: Evidence from morphological and genetic analyses." *SIDA, Contributions to Botany* 21 (2): 683–692.

Salvo, E. L. 1984. "A statistical description of nutrient flux rates of the Patuxent River in Maryland." Master's thesis, Department of Biostatistics, The Johns Hopkins University School of Hygiene and Public Health.

Schell, J. 1982. *The Fate of the Earth*. Alfred A. Knopf, New York, NY.

Schumer, M., and Jin, E. J. 2006. "Eggshell removal in shorebirds." Biology 242, fall, https: //www.reed.edu/biology/professors/srenn/pages/teaching/web_2006/MollyJen niferWebsite_20061205/references.html.

Scott, D. 1998. "A breeding congress." *Natural History* 107 (8): 26–28.

Shomette, D. G. 1995. *Tidewater Time Capsule: History beneath the Patuxent.* Tidewater, Centreville, MD.

Sipple, W. S. 1999. *Days Afield: Exploring Wetlands in the Chesapeake Bay Region.* Gateway Press, Baltimore, MD.

Skeat, W. S. 1968. *An Etymological Dictionary of the English Language.* Clarendon Press, Oxford, UK.

Smith, S. B., McPherson, K. M., Backer, J. M., et al. 2007. "Fruit quality and consumption by songbirds during autumn migration." *Wilson Journal of Ornithology* 119 (3): 419–428. https://doi.org/10.1676/06-073.1/.

Sommerfield, R. A., and La Chapelle, E. 1970. "The classification of snow metamorphism." *Journal of Glaciology* 9 (55): 3–17.

Stephenson, T., and Whittle, S. 2013. *The Warbler Guide.* Princeton University Press, Princeton, NJ.

Stockton, S. A., Allombert, S., Gaston, A. J., and Martin, J.-L. 2005. "A natural experiment on the effects of high deer densities on the native flora of coastal temperate rain forests." *Biological Conservation* 126 (1): 118–128.

Stoddard, M. C., Hou Yong, E., Akkaynak, D., et al. 2017. "Avian egg shape: Form, function, and evolution." *Science* 356 (6344): 1249–1254. doi:10.1126/science.aaj1945.

Sueur, J., Mackie, D., and Windmill, J. F. C. 2011. "So small, so loud: Extremely high sound pressure level from a pygmy aquatic insect (Corixidae, Micronectinae)." *PLoS ONE* 6 (6): e21089. https://doi.org/10.1371/journal.pone.0021089/.

Sundberg, S. 2010. "Size matters for violent discharge height and settling speed of *Sphagnum* spores: Important attributes for dispersal potential." *Annals of Botany* 105 (2): 291–300. doi:10.1093/aob/mcp288. Also available online at PubMed, PMC 2814761, PMID 20123930.

Swarth, C. W. 2010. "Beaver ponds: Bounty and benefits." *Marsh Notes* 24 (2): 1–2, 5, 8.

Swarth, C. W., and Kiviat, E. 2009. "Animal communities in North American tidal freshwater wetlands." Pp. 71–88 in *Tidal Freshwater Wetlands*, ed. A. Barendregt, D. F. Whigham, and A. H. Baldwin. Backhuys, Leiden, Netherlands.

Sweeney, B. W., and Vannote, R. L. 1982. "Population synchrony in mayflies: A predator satiation hypothesis." *Evolution* 36 (4): 810–821. doi:10.2307/2407894.

Tallamy, D. W. 2009. *Bringing Nature Home: How You Can Sustain Wildlife with Native Plants.* Timber Press, Portland, OR.

Tangley, L. 2017. "Bringing back the light." *National Wildlife,* 30 May, https://www.nwf.org /Home/Magazines/National-Wildlife/2017/June-July/Conservation/Firefly-Decline/.

Taylor, J. W. 1992. *Birds of the Chesapeake Bay*. Johns Hopkins University Press, Baltimore, MD.

———. 1998. *Chesapeake Spring*. Johns Hopkins University Press, Baltimore, MD.

Teale, E. W. 1951. *North with the Spring*. Dodd, Mead, New York, NY.

———. 1953. *Circle of the Seasons*. Dodd, Mead, New York, NY.

———. 1978. *A Walk through the Seasons*. Dodd, Mead, New York, NY.

Thoreau, H. D. 1862. "Autumnal tints." *Atlantic Monthly*, October.

———. 1906. *Journals*, 14 volumes, ed. Bradford Torrey. Houghton Mifflin, Boston, MA.

Thorington, K. K. 1999. "Pollination and fruiting success in the eastern skunk cabbage *Symplocarpus foetidus* (L.) Salisb. ex Nutt." *Journal of Biospheric Science* 2 (1). https://www.mtholyoke.edu/courses/mmcmenam/journal.html.

Tilman, N. 2009. *The Chesapeake Watershed: A Sense of Place*. Chesapeake Book Company, Baltimore, MD.

Tiner, R. W., and Burke, D. G. 1995. *Wetlands of Maryland*. US Fish and Wildlife Service, Ecological Region 5, Hadley, MA; and Maryland Department of Natural Resources, Annapolis, MD.

Tinker, D. 2014. "Dead trees are anything but dead." *National Wildlife Federation Blog*, 16 July, blog.nwf.org/2014/07/dead-logs-are-anything-but-dead/.

Tunnicliffe, C. F. 1946. *Bird Portraiture*. The Studio, London, UK.

US Forest Service. 1988. *From the Forest to the Sea: The Story of Fallen Trees*. General Technical Report PNWGTR229. Pacific Northwest Research Station, US Department of Agriculture, Portland, OR.

———. [multiple dates]. "Reports, publications, and other resources." National Lichens & Air Quality Database and Clearinghouse, gis.nacse.org/lichenair/?page=reports/.

War, A. R., Paulraj, M. G., Ahamd, T., et al. 2012. "Mechanisms of plant defense against insect herbivores." *Plant & Signaling Behavior* 7 (10): 1306–1320.

Wedell, N., Wiklund, C., and Bergstrom, J. 2009. "Coevolution of non-fertile sperm and female receptivity in a butterfly." *Biology Letters* 5 (5): 678–681. doi:10.1098/rsbl.2009.0452. Also available online at PubMed, PMC 2781977, PMID 19640869.

Weller, M. W. 1978. "Management of freshwater marshes for wildlife." Pp. 267–284 in *Ecological Processes and Management Potential*, ed. R. E. Good, D. F. Whigham, and R. L. Simpson. Academic Press, New York, NY.

———. 1994. *Freshwater Marshes: Ecology and Wildlife Management*. University of Minnesota Press, Minneapolis, MN.

Whitaker, J. O., and Hamilton, W. J., Jr. 1998. *Mammals of the Eastern United States*, 3rd edition. Comstock, a division of Cornell University Press, Ithaca, NY.

Williamson, D. I. 2012. "Introduction to larval transfer." *Cell & Developmental Biology* 1 (6): 108. https://www.omicsonline.org/open-access/introduction-to-larval-transfer-2168-9296.1000108.php?aid=8887/.

Wilson, E. O. 1998. *Consilience: The Unity of Knowledge*. Knopf, New York, NY.

Wohlleben, Peter. 2015. *The Hidden Life of Trees: Why They Feel, How They Communicate; Discoveries from a Hidden World*. Greystone Books, Vancouver, BC.

Wojtech, M. 2011. *Bark: A Field Guide to Trees of the Northeast*. University Press of New England, Hanover, NH.

Wolchover, N. 2012. "Why does fall/autumn have two names?" LiveScience, 2 October, https://www.livescience.com/34260-fall-autumn-season-names.html.

Wotton, R. S., ed. 1994. *The Biology of Particles in Aquatic Systems*, 2nd edition. CRC Press, Boca Raton, FL.

Yahner, R. H. 2001. *Fascinating Mammals: Conservation and Ecology in the Midwestern States*. University of Pittsburgh Press, Pittsburgh, PA.

Young, K. 2010. "Where have our ground-nesting birds gone?" Fairfield County Deer Management Alliance, www.deeralliance.com/node/41/.

Zenzal, T. J., and Moore, F. R. 2016. "Stopover biology of ruby-throated hummingbirds (*Archilochus colubris*) during autumn migration." *Auk* 133 (2): 237–250.

Sources Specific to Jug Bay

Burke, J., and Swarth, C. 1997. *Tree and Shrub Habitats at Jug Bay Wetlands Sanctuary*. Technical Report of the Jug Bay Wetlands Sanctuary, Lothian, MD.

Fogel, M. L., and Tuross, N. 1999. "Transformation of plant biochemicals to geological macromolecules during early diagenesis." *Oecologia* 120 (3): 336–346.

Friebele, E. 2001. "Wild rice vanishes as resident geese multiply." *Marsh Notes* 16 (1): 1, 4–5.

Friebele, E., Swarth, C., and Stafford, K. 2001. *The Ecology and History of Jug Bay: A Volunteer's Guide*. Chesapeake Bay National Estuarine Research Reserve–Maryland; Maryland Department of Natural Resources; and Jug Bay Component of CBNERR–MD; Jug Bay Wetlands Sanctuary; and Patuxent River Park, Annapolis, MD.

Friebele, E., and Zambo, J. 2006. *A Guide to the Amphibians and Reptiles of Jug Bay*. Jug Bay Wetlands Sanctuary, Lothian, MD; and Chesapeake Bay National Estuarine Research Reserve–Maryland, Annapolis, MD.

Greene, S. E. 2005. "Measurements of denitrification in aquatic ecosystems: Literature review and data report." University of Maryland Center for Environmental Science, Chesapeake Biological Laboratory, Solomons, MD.

Hentati, Y. 2016. "White-tailed deer (*Odocoileus virginianus*) monitoring and population density at Jug Bay Wetland Sanctuary." Friends of Jug Bay Summer Internship, October. jugbay.org/files/uploads/docs/FinalReport%20Hentati_Nov16.pdf.

Marchand, M. 1998. *Population Ecology and Movement Patterns of the Eastern Box Turtle*. Technical Report of the Jug Bay Wetlands Sanctuary, Lothian, MD.

Marchand, M., Quinlan, M., and Swarth, C. W. 2004. "Movement patterns and habitat use of eastern box turtles at the Jug Bay Wetland Sanctuary, Maryland." Pp. 55–62 in

Conservation and Ecology of Turtles of the Mid-Atlantic Region, ed. C. W. Swarth, W. M. Roosenburg, and E. Kiviat. Bibliomania, Salt Lake City, UT.

Meanley, B. 1996. *The Patuxent River Wild Rice Marsh*. Maryland-National Capital Park and Planning Commission, Prince George's County, MD.

Molines, K. 2008. "Development of an identification process using digital photographs: Marbled salamander migration study." Maryland Department of Natural Resources, http://dnr.maryland.gov/waters/cbnerr/Documents/publications/Molines_Sala manderReport_JugBay_2008.pdf.

Parks, M. 1998. *The Aquatic and Terrestrial Ecology of the Red-bellied Turtle* Pseudemys rubriventris, *in the Patuxent River, Maryland*. Technical Report of the Jug Bay Wetlands Sanctuary, Lothian, MD.

Stafford, K. 1990. *A Volunteer's Guide to Jug Bay*. Chesapeake Bay National Estuarine Reserve–Maryland; Maryland Department of Natural Resources; and Friends of Jug Bay, Annapolis, MD.

Swarth, C. W. 1994. "Nutrient cycling in Jug Bay marshes." *Marsh Notes* 9: 3–5.

———. 2004. "Natural history and reproductive biology of the red-bellied turtle (*Pseudemys rubriventris*)." Pp. 73–84 in *Conservation and Ecology of Turtles of the Mid-Atlantic Region*, ed. C. W. Swarth, W. M. Roosenburg, and E. Kiviat. Bibliomania, Salt Lake City, UT.

———. 2012. *A Long Look at Jug Bay*. Castle Knob, [US].

Swarth, C. W., Delgado, P., and Whigham, D. 2012. "Vegetation dynamics in a tidal freshwater wetland: A long-term study at differing scales." *Estuaries and Coasts* 36 (3): 559–574. doi:10.1007/s12237-012-9568-x.

Teliak, A., and Swarth, C. 2016. *Findings of the Monitoring Avian Productivity and Survivorship (MAPS) Songbird Netting Study at the Jug Bay Wetlands Sanctuary: 25 Years of Monitoring Data*. Technical Report of the Jug Bay Wetlands Sanctuary, Lothian, MD.

Williams, A. 1975. *Otto Mears Goes East: The Chesapeake Beach Railway*. Meridian Sun Press, Alexandria, VA.

Wondolowski, Lora. 2002. "Diurnal activity patterns of wintering gulls at Jug Bay." Master's thesis, Bard College.

Index